# Climbing to the Top

# Climbing to the Top

Anne Ashburn Pike

First Edition 1996
Designed by Anne Ashburn Pike
Library of Congress Card Catalog Number 96-70176

ISBN 0-9653914-0-X

Printed in the United States of America
by Jostens, Winston-Salem, N.C.

Floyd S. Pike Electrical Contractors, Inc.
P.O. Box 868
Mount Airy, North Carolina
27030-0868

# Contents

# INTRODUCTION

Electrical power lines are so much a part of the American landscape that hardly anyone gives them a second thought. Utility poles and power lines are rarely noticed unless our daily routines are disrupted by a power outage. Suddenly, we become aware of the comforts and the conveniences that electricity makes possible.

It is the lineman who enables us to "flip the switch" that brings electricity into our homes and businesses. To work as a lineman, a person must be intelligent, strong, and brave enough to "climb to the top" of a pole to build and repair electrically charged lines. The dangers confronted by a lineman cannot be overstated.

As a young man, my father, Floyd S. Pike, worked as a lineman and as an inspector of line construction. He believed that he could build, maintain, and restore power lines better than anyone else. Relying on his experience and the managerial skills of my mother, Opal Forrest Pike, my parents obtained the necessary funds and located the right people to realize their dream of establishing an electrical construction company. *Climbing to the Top* chronicles the first fifty years of challenge and achievement of Floyd S. Pike Electrical Contractor, Inc.

To write this story, Anne Ashburn Pike conducted over fifty interviews, obtained oral histories, and sent out sixty-seven individualized, extensive questionnaires for research that covered a geographical area stretching from New York City to Taxco, Mexico.

Also, hundreds of employees, customers, and suppliers provided information through taped interviews, letters, photographs, and news articles. *Climbing to the Top* is the story of the many people who have shared, and those who continue to share, the dream, determination, and accomplishments that have made the company a recognized leader in the electrical construction industry and the largest electrical construction company in the Southeast today. From the two who founded Pike Electric, the company now employs over 3,400 individuals and is licensed to work in fifteen states.

In addition to its reputation for quality transmission, distribution, and underground construction, as well as commercial wiring, Pike is known as the "best in the business" for rapid mobilization and quality restoration following any kind of natural disaster; be it ice, snow, hurricanes or tornadoes. Pike employees are renowned for their caring and heroic service beyond the call of duty, especially during storm trouble, and Pike's proficient employees have a history of responding to a customer's needs in tension-filled situations where bravery and cool heads are essential to doing the job.

At the time of my father's death in 1975, his own life story resembled that of a folk hero. People in Patrick County, Virginia, still talk about my grandmother Kate, holding her youngest child, Floyd, at the funeral of his father, Joseph Pike.

Floyd S. Pike was only two years old, when his father died. Later on, as an adult, he did not conceal the fact that he ran away from home to escape his abusive stepfather. His early life experiences, however, only seemed to make him extraordinarily self-reliant, and amazingly, he managed to emerge from his childhood with his faith and trust in people undaunted. He thrived on hard work and in doing the job right. He inspired those around him to strive for personal success and to share the benefits of one's personal success with others. He believed in providing a good day's pay for an honest day's work.

We at Pike continue to do business based on Floyd S. Pike's values. We are most proud of the fact that our reputation is based on our honesty and integrity. Unlike many other businesses today, we value our employees who have many years of service as one of our company's greatest assets. It is the responsibility of these employees to pass on not only their expertise, but our ethics, to new personnel, many of whom are now second- and third-generation employees with Pike. Ours truly is a family business consisting of many, many families whose members successfully continue to blend fresh ideas and new ways into the fundamentals.

Climbing is not, has never been, and will never be easy. Our first fifty years were not without growing pains, but these years are marked by our attainment of important financial and operating goals and the continued strengthening of our desire to serve our customers well.

As we begin our next fifty years, above all else, we will maintain our integrity. We will continue our efforts to work hard and to manage well. We will "keep on climbing" with confidence that our strategies for continuing growth will allow us to build upon solid past results to achieve at levels that continue to exceed industry standards and averages.

*Climbing to the Top* recounts the compelling story of how Floyd S. Pike Electrical Contractor, Inc. began fifty years ago, and how it has become the company that it is today.

Joe B. Pike,
President and Chairman of the Board,
Floyd S. Pike Electrical Contractor, Inc.

# Forward

Contract electrical line work is unique by its very nature. It requires a work force that is smart, motivated, and persistent. This is the type of person who fills all the roles from the president to the last groundman hired. However, there is another group of people that is equally as important to making it all work. These folks are the wives and families of these men. Yes, we linemen ride out into the night during electrical storms, work over holidays for industrial outages, and leave for days to negotiate work, all the while knowing that someone is watching over the home front. Yet, our wives and families are often left wondering how and where we are, until the headlights finally start climbing the drive. One person who knows all of these feelings is Anne Ashburn Pike.

Anne Pike has been a part of Pike Electric for over thirty years. She has attended business meetings, fostered warm, genuine customer and employee relations, and provided a sounding board for both business and personal ideas over the years. All of these activities took place as she raised two children and obtained a master's degree in fine art. She knows very well the feeling of watching a husband struggle with work negotiations and travel extensively. She has watched her own son leave for storm work at three o'clock in the morning. Yet, she also knows the pride and loyalty they feel to their work and the company.

Because of Anne Pike's unique background and experience, I feel no one could be more qualified to write about Pike Electric's first fifty years. She has poured tremendous effort into the project to insure that the facts are accurate. She has spent literally weeks of time interviewing linemen, to learn how the work is really accomplished. She has tracked down retired power company executives all over the world to get their thoughts and recollections of dealing with Pike Electric. Anne Pike has examined thousands of old and new pictures to select those that illustrate the various types of projects Pike has completed over the years. She has interviewed Pike Electric's management, both past and present, for their remembrances and their outlook for the future.

For anyone who has been associated with Pike Electric or has ever performed contract line work on any level, I think you will find this book fascinating. It encompasses the work and experience of many dedicated individuals. And, it gives a rare glimpse into the hearts and minds of a company that few people see until the lights come on.

*—J. Eric Pike*

P.S. Thanks, Mom!

# Acknowledgements

To the hundreds of persons who are, or have been, a part of Floyd S. Pike Electrical Contractor, Inc.,: thank you for the extraordinary privilege of writing your history. Each of you has had a part in creating a legacy that is truly worth sharing.

I am especially indebted to Jean Williamson, Amelia Reeves, and Sandy King for their professional assistance and to Keith Isaacs for technical guidance. Your help with the tasks that were essential to this project equal the tremendous pleasure of working with you.

While writing this account, I have had the delightful experience of working with Mount Airy writer/author Wanda Urbanska. Even though her schedule was extraordinarily busy with her own publications, she found the time to edit and advise with expertise and enthusiasm.

To Joe Pike, Eric Pike, Margaret Pike, Opal Pike, Emmett Forrest, Jr., Reg Banner, and John Merritt: my special thanks for answering my limitless questions with patience and good humor, and also for reading again and again. Your editorial assistance was essential.

My thanks extend also to the following persons and companies who contributed to this account: Cooper Adams; Donald Anderson; James Anderson; Ralph Anderson; Glenn Baker; Nancy Banner; Reg Banner, Jr.; Zack Blackmon, Jr.; Zack Blackmon, Sr.; Jerry Boatright; Barry Bodenhamer; Douglas W. Booth; Gene Bond; Ben Brannock; John Brewington; Claude Brinkley; Shirley Brinkley; George Broderick; Robert Chappell; William Coley; Jeff Collins; J.C. "Pete" Dillon; Glenn Dobson; Duke Power Company; Kenneth Todd Drane; Jim Draughn; Guy Driver; Gollie Easter; Ernst & Young; First Citizens Bank; Fred Folger, Jr., Emmett Forrest, Jr.; F.D. Forrester; Marsha Frazier; Bobby Gammons; Larry Graham; Witchia Gravitte; Patricia Gwyn; Jim Hicks; Lewis R. "Snow" Holding; Doug Hooker; R.M. Hudgins; John Hunt; James Kelly Hutchens; Keith Isaacs; Richard Isaacs; Tommy Jarrell; "Cookie" Jessup; Frank Jessup; Roger Jones; Charles King; Sandy King; Leman Lanier; Rufus "Bo" Lankford; Floyd Lewis; Jasper Marion; Dianne Martin; Robert E. Matthews; Esther Maynor (Mrs. A.T. "Horse" Maynor); A. L McAlexander; William B. McGuire; Leslie Sams McPeak; Billy Mears; Haywood Merritt, Jr.; John Merritt; Kelly Merritt; Metropolitan Life Insurance Company; Roger Moody; Mount Airy Public Library; John Mullins; Gladys Mills Owens; Charlie Pack; B.B Parker; Joe D.Patterson; Glen Pehl; Eric Pike; Glenn Pike; James Pike; Joe Pike; Margaret Pike; Opal Pike; Doug

Perkins; Doug Porter; Marsha Ratliff; Robert "Barny" Ratliff, Jr.; Amelia Reeves; Joe Reid; Butch Riley; Leo "Whitey" Sams; Robert G. Schwartz; Seivers Studio; Herman, "Nick" Shoaf; Audie Simmons; Glenda Simmons; Richard Simmons; Robin Simmons; James N. "Jim" Smith; Strand, Skees, Jones, & Co.; Kelly Surratt; The Rev. Preston Taylor; Roy Thomas; John Vaughan; Jim Watts; Paul D. Webster, III., MD; Pete Welborn; George Wilhite; Ruth Wilhite; Jean Williamson; Jack Wilson, CPA; Akers Woods.

Without your memories, this story could not have been written.

My personal thanks to Joe Pike for your help with every aspect of this project.

*— Anne Ashburn Pike*

# THE DREAM BEGINS

*We were working outside of town that day.*
*I looked down the road and*
*saw a boy that looked like Floyd*
*riding in the back of a pickup coming right by us!*
*I said to the fellows: "That looks just like my brother— my 'little' brother—*
*there, on that truck-fender!"*
*The men on that crew were as surprised as I was!*
*The foreman could hardly believe me at first;*
*but then he laughed and said:*
*"Well, Jim, if it is your little brother,*
*go on out there and see him!"*

*—James Pike*

# 1

# Strong People Become Strong Climbers

He decided to ask at the police station if he could stay there for the night. The youngster had hitchhiked as far as Lewisburg, West Virginia, and needed a safe place to sleep. Floyd Swanson Pike had run away from home and was on his way to join his older brothers and a sister who were living and working in that area. The boy was only fourteen years old at the time, but according to all known accounts, he had already endured enough of life's tough times to explain his actions and to justify his decision to leave his home.

Floyd Pike was just two years old when his father died on the family's farm in a plowing accident in 1920. A few years later, his mother remarried; and, essentially, at that point, Floyd Pike's brief childhood was over. According to relatives and friends of the family, almost immediately, tension developed between the blond, blue-eyed youngster and his stepfather, and the conflict only escalated as the little boy grew older, an opinion supported by his fourth-grade teacher, Gladys Mills Owens, who lived on an adjoining farm. "Floyd Pike was one child in my classroom who obviously needed love," she said in a recent interview. "I cannot verify an abusive situation," she observed, "but I certainly think that the likelihood of abuse existed in that household."

The decision to leave his home and his mother could not have been a simple one for the young boy to make — especially after his mother and his stepfather had a child of their own. Despite the unhappiness that he felt, Floyd Pike loved and protected his baby sister, Lucy, from the day of her arrival, and he was delighted by her companionship. Years later as an adult, Floyd Pike explained succinctly that his stepfather had no love for him — and treated him so unkindly — that to leave home was his only recourse.

Possibly, the decision to set out on his own became easier as the family's financial situation grew worse. In the opinion of most friends and relatives, mismanagement by Floyd Pike's stepfather caused the ownership of the family farm to slip away from the boy's mother, tract by tract. Very early in his life, Floyd Pike became accustomed to earning his own money and providing his own living. For several years, he was able to set aside a small savings for his mother and support himself by catching and selling rabbits. When he ran away, he left an established route of regular customers that included the Blue Ridge Hotel in Mount Airy, North Carolina, and a number of grocery stores within a ten-mile radius of his home in Patrick County, Virginia.

As the story is told, that night in Lewisburg, someone at the station took the boy home, fed him, and gave him a bed. The following morning, Floyd Pike caught a ride in the back of a pickup truck that was going into the town of Hinton, West

Virginia, where he hoped to locate his older brother, James Pike. As luck would have it, Floyd Pike's brother was working as a lineman on the very road that the boy was traveling. No matter how the reunion occurred, it is a fact that when Floyd Pike found his brother, he had indeed found a home.

Legend has it that as Floyd Pike was on his way into Hinton, he not only had looked for his brother, but he had also surveyed the area for any job opportunities — such as there were at that time in rural West Virginia. Apparently, he noticed a gasoline station that had been closed for quite a while, immediately determined its potential, and located the owner. Pike persuaded him to re-open the station and let him become the fourteen-year-old "manager." Accuracy in the story's details can, of course, be questioned, but it is true that, with a job and a home, Floyd Pike found some order in his life at that point that lasted for the next five years or so — until his brother found a better job and left the area.

Floyd Pike was then, once again, very much on his own, and he needed to earn better wages. Living in West Virginia, one quickly learned that good money could be made working in the coal mines. At nineteen, the boy was legally too young for such a job, but, nevertheless, he was hired. It is said that with his ambition and propensity for hard work, he steadily progressed; and, in only a few months' time, he was offered a promotion. But a better position in the coal mine came with increased risk and danger. The job became available only because the previous worker had been killed.

"When I learned that Floyd was considering the offer," James Pike said when interviewed in 1993, "I pleaded with him to leave the mines and West Virginia. I urged him to come to North Carolina and get a job with the Duke Power Company as I had." Wisely, Floyd Pike took his older brother's advice and came to Mount Airy in the autumn of 1937, where he was hired as a lineman for Duke Power Company. James Pike, his wife, Nellie, and their three young children — a family unwavering in their generosity and caring for Floyd Pike — once again provided the young man a home. Auspiciously, his new address was just around the corner from a Forrest family on Rockford Street.

*Floyd Pike was two-years old when his father died.*

*Kate Pike – Floyd Pike's mother.*

*Homeplace of Floyd Pike in Patrick County, Virginia.*

*Floyd Pike in the 1940s.*

*He was a new boy in town, and all the girls*
*I knew wanted to meet him.*
*He was dating a girl across the street.*
*I thought he was really good-looking,*
*but I had not had a chance to meet him yet.*
*While we were playing croquet,*
*he came by to speak to the girl he knew.*
*I just went up to the porch*
*and sat in the swing.*
*A few days after that, I was playing croquet,*
*and Floyd came by again.*
*This time he came by to meet me!*
*—Opal Pike*

# 2

# A Brown-Eyed Beauty and a Dream

Emmett Forrest, Sr. and his wife, Margaret, had lived in Mount Airy only a few years at the time Floyd Pike arrived. The family had been living in Oxford, North Carolina, when their oldest child, Opal, was born. Later, the family moved to Cramerton, North Carolina where they lived until Opal Forrest completed the tenth grade. Tall and beautiful, with auburn hair and sparkling brown eyes, Opal Forrest had a happy childhood, and she was both a good student and a good athlete. Before her family moved to Mount Airy, where she graduated from high school, she was captain of the varsity basketball team in Cramerton.

Growing up during the Depression taught Opal Pike real and unforgettable lessons about hard work and careful savings. She already had several years of work experience to her credit while still a teenager. She had picked cotton and worked in a department store. She even managed an ice cream parlor for a while. At the time she met Floyd Pike, she was employed by a local textile mill in Mount Airy.

A strikingly handsome couple, each had found a partner for life. Floyd Pike and Opal Forrest were full of energy, fun, and love and respect for each other. They were truly equals in determination, tenacity, and sacrifice. They also shared a willingness to take tremendous risks. Their loyalty to friends and family and their trust and belief in people were as genuine as their faith in God. Having known each other only three months, they married December 10, 1937.

Glenn Pike, Pike Electric vice president, Commercial Division, retired, is the son of James Pike. He was still a young child at the time Floyd Pike lived with his family, but he clearly remembered his uncle's wedding day. "Floyd was standing before his mirror, humming, then singing," Glenn Pike said. "I watched and wondered, and I finally asked, 'What in the world is the matter with you?'" Floyd Pike laughed and quickly explained, "Opal and I are getting married today!" Glenn Pike remarked, "He seemed to be the happiest man I'd ever seen."

At the time of his marriage, Floyd Pike had been promoted from a lineman to an inspector of power line construction for Duke Power. The power company's local manager in Mount Airy then was R.M. Hudgins. Floyd Pike's acquaintance with his employer developed into a friendship and mentorship that lasted throughout Pike's life.

During those early days of employment, another of Floyd Pike's pivotal and lasting friendships began when he met H. E. Carter, who coordinated and oversaw the power line construction done by the electrical contractors that served Duke Power Company. In the course of a day's work, H.E. Carter was often on the job site with the young employee. As Floyd Pike inspected the line work of contractors who, only a few years' later, became Pike's competitors, he, astutely and without reservation, pointed out to his boss any construction he deemed unsatisfactory. Characteristically, Floyd Pike's observations ignited his imagination, and the possibility of a business began to form. While he worked and watched, he became convinced that he could certainly build power lines better than any of the construction he saw being done.

This was the point in time when the Pike's dream began. He was only twenty years old.

Although, with World War II approaching, circumstances were uncertain for everyone, this young man began to believe that someday he would have an electrical construction business of his own. In an article, "Know Your Neighbor," in *The Mount Airy News*, January 3, 1958, reporter Bob Tannehill, interviewed Pike. "I knew then what I wanted," Floyd Pike said, "and I waited and watched for the right opportunities."

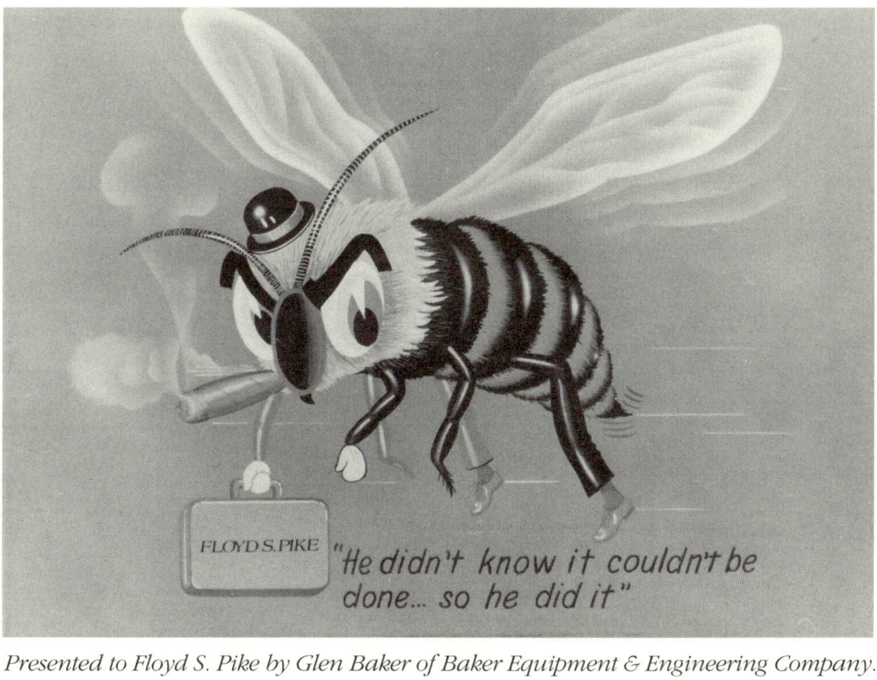


*Presented to Floyd S. Pike by Glen Baker of Baker Equipment & Engineering Company.*

*The Pikes in 1938 soon after their wedding.*

*Floyd and Opal Pike – 1940.*

*I knew then what I wanted.*
*— Floyd Pike*

*We had a neighbor build us a little one-room trailer.*
*It wasn't very big; I'd guess twelve or fifteen feet,*
*or smaller than that. It had only one door that*
*opened onto a little back porch.*
*Floyd had been working at Cherry Point,*
*and he came home to take me — and the trailer — back with him.*
*He was so tired. He said that I should drive for a while,*
*and he would sleep back there in the trailer.*
*Later, he would drive the rest of the way.*
*After a few hours, I pulled off to stop.*
*I walked toward the back of the trailer,*
*thinking Floyd would still be asleep.*
*But he wasn't! He was already out*
*and coming around the side to meet me!*
*I said to him, "Well, I thought maybe I'd have*
*to come back and wake you up."*
*"Wake me up! Nothing!" he said.*
*"I haven't slept any all night! I've been running up*
*and down the highway, trying to catch you!*
*Every time we came through a town, and you would slow down,*
*I'd try to get to the car. I'd get almost close enough*
*to get a-hold of the door handle;*
*and you'd pull off again.*
*Then I'd have to run and jump on that trailer-porch,*
*so I didn't get left out on the road somewhere!"*
*—Opal Pike*

# 3

# Days at Cherry Point

War preparations on the North Carolina coast in the early '40s included construction of the Cherry Point Marine Corps Air Station, near Camp Lejeune, North Carolina. Among the linemen coming from work at the base, conversation was filled with exclamations about plentiful defense-related jobs, with wages that seemed extraordinary. The Pikes recognized an opportunity. With R.M. Hudgins's understanding and encouragement, Floyd Pike left Duke Power to work for Thompson Electric, a contractor responsible for the electrical construction at the air base.

In their move to Cherry Point, the Pikes joined hundreds of other persons arriving with their temporary housing in tow. And if living in a military area during the war and working on the military post itself were not exciting and unsettling enough for the young couple, Floyd Pike, like most of the nation's young men, also faced the wartime draft. Together with many of his friends and acquaintances, he joined the Navy.

Having just completed his basic training, Pike returned to Cherry Point on short leave. On one night of a fortuitous weekend, a squadron of planes was on approach and scheduled for landing at the base. But because of a fire in a bank of transformers, the runway-lights were out. With war shortages, there were no replacement transformers available. Right away, news of the difficulty spread.

When Floyd Pike learned of the problem, he hurried to the site. Expeditiously, he opened the system, made the necessary repairs, and lighted the airport. Almost immediately, he was recognized for his voluntary help, and the post commander quickly issued orders to transfer Pike to the Merchant Marines in order to keep the young electrician at the air base in charge of electrical maintenance. Even after the war ended, Floyd Pike's responsibilities continued, although his employer was then the Civil Service, so the Pikes remained at Cherry Point several months longer, albeit, in a "new, store-bought" trailer. Opal Pike explained that they had lived in the tiny, hand-built trailer only a short time. "We thought it was nice until we got down there and saw all those pretty, new ones. Then we bought us a National. It was wonderful, with so many conveniences !"

*"Housing in tow." Floyd and Opal Pike's trailer at Cherry Point, North Carolina 4th from left.*

*Days at Cherry Point, North Carolina.*

*Lineman Floyd Pike, left.*

*Floyd, Opal and Joe Pike.*

*Floyd Pike and Joe Pike at High Rock Lake in 1945.*

*Lineman Floyd Pike at Cherry Point.*

*This Mr.Thompson thought Floyd was a good lineman*
*and a good supervisor.*
*He offered him a really good deal to leave Cherry Point*
*and go to Raleigh, where Thompson Electric was based.*
*He promised Floyd one line crew.*
*If that one crew made a profit in six months or so,*
*Mr.Thompson said we'd add another one and another one;*
*and then we'd just keep on growing.*
*He promised us a small salary, but, more importantly,*
*he promised us a percentage of the profit.*
*I kept the books — all of the records of the payroll,*
*the materials that we had to have to build that line,*
*and the amount of money that was collected.*
*Floyd was so anxious to make a profit;*
*he would work really long hours.*
*Then, at night, we'd go out to the place*
*where the trucks were parked and load the materials.*
*Our son, Buddy,(Joe Pike) was really young.*
*We'd give him insulators to play with,*
*while we loaded the trucks.*
*But we really weren't thinking about it at the time;*
*you see, we were getting an education.*
*Then we were so disappointed!*
*The promises ... well, we just didn't feel that they*
*matched the paycheck.*
*I don't know... It could have been,*
*maybe, we just didn't know something about the overhead.*
*It has turned out well for many people that we felt that way.*
*—Opal Pike*

4

# The Climb Begins

The Pikes were understandably disillusioned by their experience with Thompson Electric. However, when they assessed all that they had learned, they began to feel quite sure of their capabilities. Their confidence produced enough determination to spark their optimism. Floyd and Opal Pike decided then to leave Raleigh and return to Mount Airy. With their two-year-old son, Joe, a small savings, and a big goal in mind, they decided "to take a chance" on an electrical construction business of their own.

The Pikes' decision was a timely one: there was work waiting to be done. Joe Maynor, author of *Duke Power: The First 75 Years*, described the work-related environment of 1945:

> Like most other private industries around the country, the immediate post-war era for Duke Power Company was a time for trying to play catch-up. Lines and equipment had been laboring under four years of strain. During the war years, materials and supplies were virtually unavailable to private industry, except those fundamental things necessary to continue services vital to the war effort. As a result, Duke Power's facilities had been spliced and patched and pressed back into service time and time again. Now that the war was over, company officials expected a heavy demand of electricity from increasing industrial development and untold thousands of new citizens in the Piedmont Carolinas. Lines and equipment that had deteriorated over the years would not stand up to the demand.[1]

As soon as he got his family settled, Floyd Pike went to Charlotte to meet with H.E. Carter. The ambition, knowledge, and integrity that Floyd Pike had displayed when he had worked under Carter years before were exactly what the young man offered in exchange for a contract. Having only the small amount of money they had managed to save, it was vital for the Pikes to obtain a contract that provided pay for work as it was completed, along with the materiel required for the job. Again, Opal Pike was designated bookkeeper.

Writing to Pike in March 1993, R.M. Hudgins, now retired after many years with Duke Power, remembered: "H. E. Carter called me and wanted to know if I thought Floyd Pike could do the job. I told him, 'Yes'!" By all known accounts, Hudgins response only confirmed what H.E. Carter already believed.

Unpretentiously, the little company was simply called, Floyd S. Pike, Electrical Contractor. With one contract, for one crew to provide service to one customer — Duke Power Company

— the beginning occurred. Whoever then would have dreamed that this working relationship would be continuous fully fifty-years later, when Duke's major power line contractor would be Floyd S. Pike Electrical Contractor, Inc.?

Without a doubt, Floyd and Opal Pike!

*Floyd Pike and H.E. Carter.*

*Somehow, the truck had toppled off*
*a barge into the water between*
*Ocracoke Island and Morehead City, North Carolina.*
*When it was recovered the damage included*
*a cab with no top, no side-boxes,*
*and most of the windshield was missing!*
*Floyd and his cousin, Billy Mears,*
*were driving the thing home to Mount Airy.*
*I waited up all night....*
*It was winter— November— and so cold.*
*I remember that when Floyd came in,*
*he was almost frozen!*
*But he was about the happiest person I have ever seen.*
*Laughing and talking, he said to me:*
*"Opal, we've got us a contract;*
*and now we've got us a truck!"*
*We declared ourselves to be in business,*
*December 3, 1945!*

*—Opal Pike*

5

# Getting Started

The young couple began to deal with the essentials for getting the job done. Bottom line, they needed capital. Added to the funds that Floyd and Opal Pike had managed to save during the first few years of their marriage, Floyd Pike convinced his cousin, Billy Mears, to invest a small amount of money that he had made during the war years, while working at a draft-deferred job with a dairy not far from the Pike farm. With Billy Mears as a working partner, Floyd Pike requested, then refused, much needed additional funds from a local businessman in Mount Airy, who required not only a partnership, but also greater interest on a loan than the Pikes felt they could assume.

*Bill Mears and Floyd Pike.*

Consequently, Floyd Pike went home to Patrick County, Virginia and sought financial help from a lifelong friend. With generosity that is typical of the residents in that area, Ramey Vaughn, a neighboring farmer, provided the additional loan necessary for business to begin — interest-free and repayment unquestioned. Billy Mears recalled that the amount of money Vaughn pulled from the bib-pocket of his overalls was $5,000.

With funds in hand, the young entrepreneurs began to deal with other necessities. Along with the work experience gained at Cherry Point, the time there also proved resourceful for locating equipment and for hiring people. With the post-war shortages, equipment was extraordinarily scarce, not only for individuals who were just starting out, but for large companies as well. *Duke Power Company: The First 75 Years* corroborated the difficulties: "Although the war was over, there was still a severe shortage of materials. Only an expert at procurement could locate and buy the unique materials that had to be used in replacing electric power lines."[2]

*Billy Mears in 1945.*

From all people and places to search for essential equipment, Pike's pair of "expert[s] at procurement" contacted Fred Davis, manager of the commissary at Cherry Point, and from this reliable, albeit irregular, source, Pike and Mears learned of an available line truck for sale.

Unique? Unquestionably! The line truck was actually underwater at the time. With a purchase price of $250, the company's first piece of equipment was literally pulled from the ocean near Beaufort, North Carolina. That line truck, now painstakingly restored by Pike Electric Field Mechanic, Chuck Fischel, an automotive restoration specialist, is still proudly cherished by the company and is fondly and respectfully referred to as "Old No. 1."

To equip the truck, Floyd Pike and his partner paid another $250 for one pole-lifting winch. In addition, Billy Mears said that the partners were able to locate and buy a used three-quarter-ton '37 Ford pickup truck, and a "bunch of ropes and some pulleys."

"New" equipment ... contract in hand ... ready to go to work: all that was lacking was a crew!

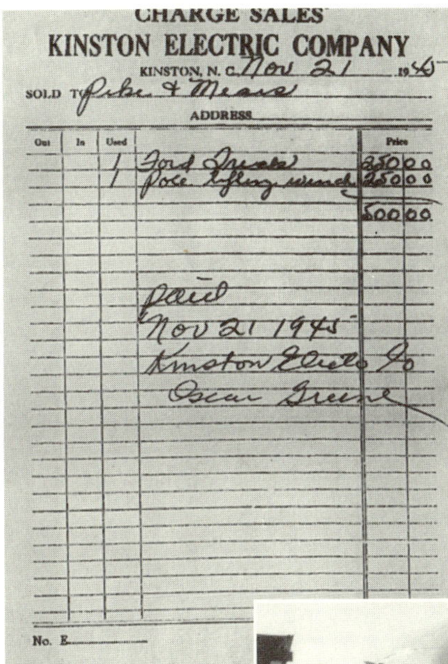
CHARGE SALES
KINSTON ELECTRIC COMPANY
KINSTON, N. C. Nov 21 1945
SOLD TO Pike & Mears
ADDRESS

| Out | In | Used | | Price |
|---|---|---|---|---|
| | | 1 | Ford Truck | 250 00 |
| | | 1 | Pole lifting winch | 250 00 |
| | | | | 500 00 |

Paid
Nov 21 1945
Kinston Elect Co
Oscar Greene

No. E

Bill of Sale for
"Old Number One".

"Old Number One".

Opal Pike and Chuck Fishel in restored
"Old Number One".

*Floyd was a determined kind of man.*
*Like the time he got after me*
*to come to work for him. I put up all kinds of excuses,*
*and he would always head me off.*
*So it finally came down to*
*I didn't have any boots, and you couldn't buy `em in those days.*
*Floyd said, "That is no excuse;*
*I've got a brand new pair that I'll give you."*
*So I had to go to work then. And I've never been sorry.*
*You have heard of a man giving the "shirt off his back"?*
*Well, Floyd would give the shoes off his feet!*
*He was that kind of a boy. No kidding!*
*He put a meal ticket in the hand of many a man*
*who might not ever have had one otherwise.*

*—J.C. "Pete" Dillon*

# 6

# The First Climb

Not only was there a shortage of equipment at the war's end, linemen were scarce as well. Along with Floyd Pike and Billy Mears, the first Pike crew included: J.C. "Pete" Dillon, J.C. Harris, Fred P. Howell, Glenn Pruitt, Carl Scott, and Edgar McMillan. By the second week, Bob Belton, Gilmer Reece, Cecil McMillan and Edd Thomas were on the payroll. Billy Mears recalled that top linemen were paid $1.25 an hour.

Help in hiring adding additional men was provided by the GI apprentice program. This governmental assistance granted supplementary pay to returning servicemen while they retrained for civilian work. During the first few weeks that Pike was in operation, James Anderson, now retired after more than forty years of service, said that when he came to work in February '46 and he was paid 45¢ an hour, plus the government's supplementary pay: "I had been in service, and when I came home, Opal helped me get the GI Money." The lineman-program lasted for four years and was beneficial to both employer and employee. "We had a lot of people come to work under the GI Training program," Anderson added, "but many of them could progress faster than the GI scheduled pay and would bypass the use of it."

The first crew began its work for Duke Power Company by pulling wire down Cherry Street in Mount Airy. From R.M. Hudgins's descriptive letter, one can imagine the little crew on its first endeavor : "The next day material arrived, and Floyd arrived with his crew and a two- or three-ton flatbed truck. I believe this was the first contract job Floyd performed. He did an excellent job for us." Claude Brinkley, Pike Electric vice-president, Northern Division, retired, said that everything did not always go quite as planned for that first crew.

> I couldn't help but chuckle when I heard this story, but I'm sure at the time that it happened, it wasn't funny. During the years that my dad was the local power company superintendent, he worked closely with all the Pike crews in the Mount Airy area. It seems the first crew had been working on that first contract job for several weeks and had been making good progress.
>
> One day, the crew gathered, as usual, at the warehouse and went out to work. Sometime during the morning, Dad went to the job site to check to see how everything was going. When he drove up, he saw all the men standing around the truck with puzzled looks on their faces. When Dad asked them what the problem was, the foreman explained that they couldn't work without a brace and a bit to bore holes. And the

only lineman in the crew who had one, had quit the day before. Since no one in the crew had any money to buy one, Dad loaned them the braces he had in his truck until someone could afford to buy the equipment they needed.

Starting out was not easy!

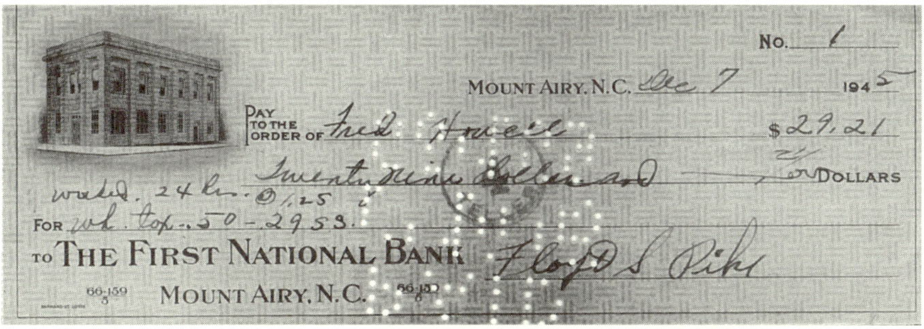


*First paycheck issued by Floyd S. Pike Electrical Contractor to Fred Howell.*

*Floyd Pike, T.E. Brinkley (father of Claude Brinkley), J. Ed Brannock (father of Ben Brannock).*

*If Floyd profits,*
*then we will profit.*
*—"Horse" Maynor*

# 7

# We Called Him "Horse!"

In the quest to find qualified linemen that first year, an unforgettable lineman found the Pikes. One day, most unexpectedly, Opal Pike received a call from the employment office inquiring if she knew anyone by the name of A. T. Maynor. Opal Pike asked if the man's name was "Horse."

Somewhat puzzled, the caller responded: "Mrs. Pike, do you mean 'Horace'?" "No. I mean, 'Horse'!" Opal Pike said that she heard his familiar laugh, when Maynor answered, "'Horse!' Oh, yes!... That's me!"

Floyd Pike and A.T. Maynor had been "pole-partners" at Cherry Point while they worked for Thompson Electric. Maynor had just been discharged from service after the war, when he learned about the young Pike company through one of its new employees, Berkley Snider, who also had worked for Thompson at the base. Maynor's wife, Esther, recalled that she and her husband "traveled by bus to Mount Airy in February, '46, and we checked into the Blue Ridge Hotel on a Sunday night." Maynor began working for Pike one week later.

A.T. Maynor — called "Horse" by almost everyone who knew him — became legendary with the Pike company. Only three months older than his employer, he and Floyd Pike could not have been closer, if brothers. This opinion was supported by Esther Maynor when she recalled the time her husband was in the hospital following his first heart attack. "Floyd Pike was a continual visitor," she said, and "he was very concerned about our financial situation because we had just begun building a new house." Although she assured Floyd Pike that they were financially secure, nevertheless, a bank deposit slip arrived in the mail, labeled "just in case!" Maynor was not told about the surprise deposit until he recovered, and than they returned the money. "This is only one example of Floyd Pike"s generosity," Esther Maynor said, "but it remains an unforgettable, important memory."

Remembering her friend and co-worker, Opal Pike described "Horse" Maynor as fair and honest. "He was really handsome, tall and thin, with very black hair, and a little mustache," He had the nicest smile and the warmest personality — got along great with people. He was really a wonderful person." She added that he could be "strict with those boys" on the job, but that they really respected him. "They knew that when he was telling them something to be safe, he was looking after their lives. 'Horse' was a very valuable man."

Emmett Forrest, Pike Electric vice-president, secretary, and treasurer, retired, called "Horse" Maynor "a really, really important cog, early in the company." Forrest said that Maynor's method of training, especially with the field personnel, was firm, but respectful. "He never raised his voice. He had a calm about him, at all times, under all circumstances.

'Horse' had the respect of everybody. If he got mad, it didn't show." Forrest added, "I have had field employees tell me: 'I'd rather be chewed-out by 'Horse,' than for some people to call me in and give me a raise.' He didn't belittle the men, he corrected them."

Emmett Forrest's words preface well the memories of Mount Airy attorney Fred Folger, Jr.:

> I worked one, maybe two, summers for Pike Electric as a right-of-way cutter and a hole-digger. My crew chief or supervisor was "Hoss" Maynor, who was one fine gentleman and an excellent supervisor. For the first hole I dug, "Hoss" left me in the middle of a pasture. Between the bull in the pasture, around which I walked carefully, and the hole I was to dig, it took much longer than "Hoss" thought it should have. I, not knowing exactly the required depth and diameter of the hole, had gone much too deep and much too big in diameter. "Hoss's" only remark was, "Now I see why it took you so long. Let's go dig another hole."

Joe Pike, Pike Electric president and chairman of the board since '75, recalled the wonderful times he had "not only growing-up with 'Horse' as my friend, but, along with Esther, they were really like part of our family." He said that "Horse" was like another father, and Joe Pike emphasized how valuable the hours and miles he spent in training with Maynor were, and how essential the training has continued to be. "Horse" had the most amazing ability to figure any job," Pike said. "He literally could drive down a road where a line was proposed, and figure the cost of the job in question by the time he reached his destination, mentally determining the number of men, number of poles, the time and equipment necessary for a job."

Esther Maynor, in the tradition of support that Pikes' wives and families are known for, often rode along with her husband when he set out to bid a job. "One day, when he had to bid a job in Charlotte, he handed me a typewriter and told me to get in the back seat of the car," she said. "On the way there, I called out each segment of the bid, and he told me what to type." She said that when they we got to Charlotte and the bids were opened, that "Pike's bid was so close to the engineer's estimate that they thought someone had called Maynor and told him what to bid!"

Joe Pike added that, whenever necessary, "Horse" Maynor could devise quite clever methods to get a job done. An example of Maynor's ingenuity was a memorable job for the Carolina Power and Light Company in the '50s. Roy Thomas, Pike Electric vice-president, Southern Division, retired, was the foreman on the job and well remembered the details:

> Because the job was expected to last about a year, I moved my wife, Mary, and our baby, Lynn, from where I had been working in West Virginia to Swansboro, North Carolina. The job involved rebuilding and re-conductoring a 115,000V line approximately thirty miles long. We worked from a point near Jacksonville, North Carolina, through the Croatan National Forest, to intersect with an existing transmission line near New Bern, North Carolina. One by one, new poles were set. In order to keep this line "hot" while we worked, we attached temporary fifteen-foot cross-arms at 45 degrees, above the existing crossarm on the H-structure. This carried the load; while, mile after mile, of new 795MCM/ aluminum wire was pulled to replace 2/0 copper. This was done for all three new wires. With a couple of 115K switches, one with a ring, and the other with a blade, we simply "blew the fuse" by pulling the ring with a hot-stick! We did this to de-energize each time it was necessary for transferring the load from one conductor to the other, letting the blade energize both the new and the temporary. At each conductor, we installed permanent jumpers and "blew the fuse" again to remove the temporary conductors. Then we moved on, and did it all, over and over again.

Respect for this gentleman and recognition of "Horse" Maynor's capabilities extended far beyond the Pike employees. After he moved from the field into management, much of his time and attention was spent directly with Pike's customers, in particular, the executives and supervisory personnel of the utilities, municipals, and cooperatives for whom the company works. A result of his love of fishing was an annual trip to Morehead City that the company hosted for many years. The fishing trips became widely anticipated events for customers and employees alike.

*Annual fishing trip hosted by Pike. Horse Maynor (top, back).*

The tradition of hunting, fishing, and golfing trips that Floyd Pike and "Horse" Maynor began in the '50s has been shared, enjoyed, and continued through five decades. These outings were opportunities to reiterate a tenet that Floyd Pike strongly believed. He often said, "When our customers get to know us, they will know that our service and integrity are as sure and certain as our friendship." The company has always practiced what Floyd Pike preached. "If you can't do business as friends, there is little hope for doing business otherwise." This philosophy has provided Pikes' people with many personal friendships among its customers and suppliers, as well as within the company itself — friendships and life experiences that continue to exist far past office hours or an individual's years of employment. And Pike's man called "Horse" was, above all else, a friend.

In a letter to Opal Pike, H.E. "Nick" Shoaf, now retired from Duke Power's senior management, summarized: "'Horse' Maynor came along, and we developed a close friendship. He had an uncanny ability to visualize how work could be done quickly, efficiently, and, above all, safely."

"Horse" Maynor's life and career were, in fact, far too short. "He was in Charlotte to play golf with Duke Power vice-president Pat Huff, and he had planned to spend the night there. Before daylight, he became ill and checked-out at 3:00 a.m. In an attempt to reach his doctor, he drove to Baptist Hospital in Winston-Salem, North Carolina where he died in the emergency room on June 7, 1968," Esther Maynor poignantly concluded.

"Horse" Maynor came to work, first, as a lineman. Through the years, he trained hundreds of other linemen, including many who comprised the nucleus of Pike Electric's senior management and supervisory personnel. Some are retired; others remain, continuing and committed to lessons learned from this man, now twenty-eight years after his death. All who worked with him learned, by example, not to flinch. With the employment of "Horse" Maynor, Floyd Pike began building a management team with qualities of excellence that remain, to this day, standards for the company.

# Pike Meet Points To Company Growth

Saturday evening at Blu-Vue Restaurant here, 69 stockholders of Floyd S. Pike Electrical Contractors, Inc., heard the record of company growth outlined. The company's annual volume of business is currently 16 times greater than it was when the company was incorporated in 1953.

Floyd S. Pike, president, welcomed the stockholders, all company employees, from points as far away as Baltimore, Md., and New Orleans, La. There are a total of 100 stockholders.

The annual meeting also served as an occasion to report on the company's profit sharing retirement fund. Financed by the company from its profits, but owned by the employees, the fund totals approximately $900,000 after three years in existence.

Pike also paid tribute to A. T. (Horse) Maynor, who died in June, calling him "one of the best friends I ever had". Maynor was a vice president of the company at the time of his death.

Director elected for the coming year include: Floyd Pike, Opal F. Pike, Joe B. Pike, Emmett Forrest, Jr., Haywood Merritt, Jr., Zack Blackmon and Roy W. Thomas.

Meeting immediately after the stockholders meeting, the directors named the following officers:

Floyd S. Pike, president; Joe B. Pike, executive vice president; Opal F. Pike, first vice prsident; Haywood Merritt, Jr., vice president; Roy Thomas, vice president; Zack Blackmon, vice president; Glenn E. Pike, vice president; Lawrence D. Allen, assistant vice president; A. L. McAlexander, assistant vice president; Joseph Fulk, assistant vice president; Emmett Forrest, Jr., secretary and treasurer; Allan Brown, assistant secretary and treasurer.

*Press covers Pike Shareholders' meeting in 1968, the year of "Horse" Maynor's death.*

*Back near the company's beginning,*
*we had a foreman who said:*
*"This company's never 'gonna' amount to nothing!"*
*Well, the word got back to Floyd, and he fired the guy.*
*Floyd said: "If I have to,*
*I'll run the crew and let Opal be the grunt!"*
*—Pete Dillon*

# 8

# The Real Work Begins

During that first year, many other competent and dedicated men, whose names are synonymous with the company's beginnings, came onto the scene. Besides the members of the first crew, many will also remember :

Tommy Jarrell; R.W. "Shorty" Clawson; M.E. "Ed" Jessup and his brother, Gene ; the Hendricks, W.H. and R.W.; Sam Thomas; Gilmer Reece; C.W. Jones; P. R. Heaton; Berkley M. "Snider" Snider; H.B. "Howard" Inman; R.C. Hooker; J.P. Hooker; Fred Hayden; H. W. "Howard" Farmer; Sam. G. Sexton; the Willards: Prather, John, and Melvin; Harrison

Thomas; William R. Moss; the Reeces: Joe, Perry, and C.C.; Roy Pike; C.F. "Cecil" Farmer; Jack Steele; Rafe Brady; the Cookes: Chester and A.M. "Jack"; B.J. Parker; Paul Riggs; J.C. Scales; J.B. Hawks, Jr.; Nelson "Woodrow" Clement; C.E. Mayberry; G.M. Reynolds; C.E. Ledford; H.B.Inman; Wade Inman; Lester Inman; Lewis Ball; O. C. Bowman; Edgar Sealey; the Marshalls: O.V. and Walter; O'Leary Ring; Henry Green; J. B. Dollyhite ...

Before the end of the first full year, Billy Mears relinquished his partnership, but he continued to work with the company. He became a bit discouraged because the little company never seemed to be able to get ahead. "Every time we got a truck, we would put the title in the bank and borrow more money," Mears recalled. Opal Pike asked him if he wanted to sell his interest, and with her personal life-savings, she purchased his share, dissolving the partnership. "I've often thought that I should have kept some of my ownership," said Billy Mears. "I was only twenty-seven years old at the time," he added, "and as the saying goes, 'I couldn't see beyond the nose on my face!' "

*The real work begins.*

*As I was growing up, Dad was away much of the time.*
*But the times that he was home — or I was away with him —*
*were "super highs" that seemed to make up for those missing days.*
*We went to the world series once.*
*We had such terrible tickets the first day*
*that the highest ball that was pitched never got to eye level for us.*
*The next Dad got the concierge at the old Taft Hotel to work on that.*
*I have no idea what those tickets cost, but Dad and I had seats in Duke Snyder's box for World Series history: Don Larson's perfect game.*
*Dad often took me with him on business trips.*
*This bothered a truant officer in Baltimore one day,*
*when Dad was calling on Baltimore Gas & Electric,*
*and I was waiting for him alone in an arcade.*
*And there was another trip, when I was twelve years old or so, that I always remember.*
*In the '50's, we were needing a little financial help,*
*from one of the local banks in Mount Airy, so we took the bank's president with us to see some of the work that the company was doing for Appalachian Power Company in some very rough mountainous areas in Virginia.*
*We got to the job site, and were talking with the foreman,*
*when I saw Joe Fulk "snaking poles" up the side of the mountain,*
*and I headed that way.*
*Joe jumped off the dozer, and I climbed on and started to do his job for him.*
*When the banker saw me coming toward them, he was frightened speechless.*
*He could only tap Dad on the shoulder and point at me in alarm.*
*Through the years, Dad and I always had a good laugh together every time we recalled that day.*

*—Joe Pike*

9

# Home and Office

Near the end of the company's first year, Tommy Jarrell, Pike Electric supervisor, retired, came to Mount Airy: "In November 1946, there were two places that I could get a job — one was with the Pike company in Mount Airy and the other was with a telephone company that would send me to Florida. I found a Pike crew out on a job, and the foreman was Bob Belton. He told me to come in on Monday morning, and I did. In fact, I came in on Monday morning for the next fifty years!"

Tommy Jarrell certainly saw the Pike operation from its infancy – a time of hard work for everyone involved. Floyd and Opal Pike set the pace for the rest to follow. "Floyd was looking after all the crews out in the field and making customer contacts at the same time. He was doing it all." Jarrell remembered. "And Mrs. Pike was doing the office work. They had a desk in the living room of their apartment where she kept the books." In that first year, the company's tradition of doing anything required to get the job done was firmly rooted.

From its beginning and for it to succeed, the life of the company was inseparable from the lives of the Pikes individually, and as a family. "Not only for us," Joe Pike said, "but for so many members of the corporate family, the difference between home and office has really never been anything other than a line on a mailing address."

The first office was located in the Pike's apartment that was owned by C.O. King, the father of Charles "Droopy" King, presently Pike's general foreman in charge of crews serving the City of High Point Electrical System in High Point, North Carolina. In the early '50s, the office was moved out of the Pikes' apartment when the couple bought land along Spaugh Street in Mount Airy, the present location of the main office and garage facilities.

At the time of purchase, a small cinder block building and four small rental houses stood on the property, and Spaugh Street circled the area where the main office now stands. Riverside Drive, the current street address of the main office in Mount Airy, was not even in existence then. Pike personnel, Glenn Pike, Jack Farris, Ralph Hudson and Carl Norman lived with their families in the rental houses. Later, one of those houses became the first location of Pike's Commercial Wiring Division.

The side entrance area of the main office today — the designated smoking area — is the original cinder block building. In a part of the tiny space, William Mitchell, Ray Coleman, Ralph Tucker, and later his brother, Emmett Tucker, assisted Opal Pike with the bookkeeping. The remainder of the space was garage. To this first little building, additions have simply been added as needed.

It became necessary to enlarge the little block building by the mid -'50s. In the present main office building, approximately one-third of the area that is now designated the "old" conference room served as Floyd Pike's office. The front part was one big room with a picture window. This section was partitioned into small office areas for "Horse" Maynor, Zack Blackmon, Sr., and Haywood Merritt, Jr., all of whom were named vice-presidents in '56. Opposite, Emmett Forrest had a small office with a sliding window that was shared by the garage — a feature that he and A.L. McAlexander, Pike Electric assistant vice-president, retired, called their "intercom!" This construction was soon followed by a major expansion during which Pike erected a garage across the street from the main office building and more space was added to office building as well.

For decades, the office remained open at least half of the day on Saturday. Men came by to talk things over, buy supplies, and prepare for the following workweek. At Pike, from the first, no matter what day of the week or hour of the day, whatever, whenever, or wherever there was something to be done, it was. And, it is.

Sometimes "what had to be done" was surprising. For example, there was a time when Opal Pike became the company cook, an unusual job that resulted from the company's erratic work schedules. In *Duke Power Company: The First 75 Years*, Joe Maynor explains: "In order to make the required repairs at a minimum of inconvenience to customers, maintenance crews were dispatched around the system for three hours each Sunday morning. It was a time when there was a light demand for power."[3]

Opal Pike recalled that the company had only one crew then, and it was scheduled to work on a Sunday morning — a time when almost every restaurant in Mount Airy was closed. She said that Floyd Pike called and said, "Opal, there is going to be a bunch of us hungry about 12:00. Can you stir up something, so I can bring the crew home to eat?" She said that she cooked a big lunch that they all enjoyed, and then the crew went back to work.

Along with her corporate responsibilities — and her

culinary duties — Opal Pike was a working mother. "Mother ran the office with one hand, while she raised me with the other!" Joe Pike said. He added that while his mother worked, "I simply — and happily — spent my days here."

Joe Pike's responsibility and companion was his pony, "Bill". Stabled behind the office, this equine-baby-sitter, while a bit unusual, served the family well. Floyd and Joe Pike shared a love of horses from the time Joe Pike was young. Although Opal Pike did not ride, she was always present at every horse show in which father and son participated.

"With all that they doing in the early years," Joe Pike said of his parents, "they managed to keep me right in the middle of it all." He said that during the school year, his family usually had lunch together, or he walked a few blocks from school down Rockford Street to have lunch with his grandmother, Margaret Forrest. "I went to the office each day after school, and I stayed there until time to go home." That schedule continued for many years.

"One of the real advantages of growing up in the office was that everything was not all play around here," Joe Pike said. "I had all kinds of opportunities to learn how to do many different things around the office and the garage. I learned to read and draw blueprints before starting the first grade." He said that a set of hooks and a belt were standard gear for him for as long as he could remember. "Dad had me climbing poles, not long after I was walking!"

Joe Pike said that another good thing about growing up on the company grounds was learning to operate the equipment. "With all the hours I spent playing in, on, and about the company's equipment, I just automatically learned to operate every piece that the company owned before I was a teenager." He said that he had become "pretty good with a 'dozer" by the time he was ten or twelve. "I was able to 'snake' poles up a mountainside by then." Pike couldn't say when his first real job began. "I began driving a dump truck and hauling gravel for the Commercial Department — *legally* — when I was sixteen."

Although the roles of the family members blurred at times, the focus of their efforts was apparent to all. They were working hard to succeed in everything they did. Managing her home/office, Opal Pike's administrative abilities, as well as her secretarial skills, were essential to the young company. This fact did not elude the paternal eye of H.E. Carter. Opal Pike recalled that he never failed to compliment her invoices. Having questioned Floyd Pike concerning her salary, Carter suggested that she be included on the payroll. And she was. When H.E. Carter made a suggestion, it was not ignored.

*Growing up at the office.*

*Joe Pike practicing climbing skills.*

*Opal Pike, Joe Pike and "Bill".*

Floyd PIke showed his horses in the '50s and '60s.

Opal Pike on the job.

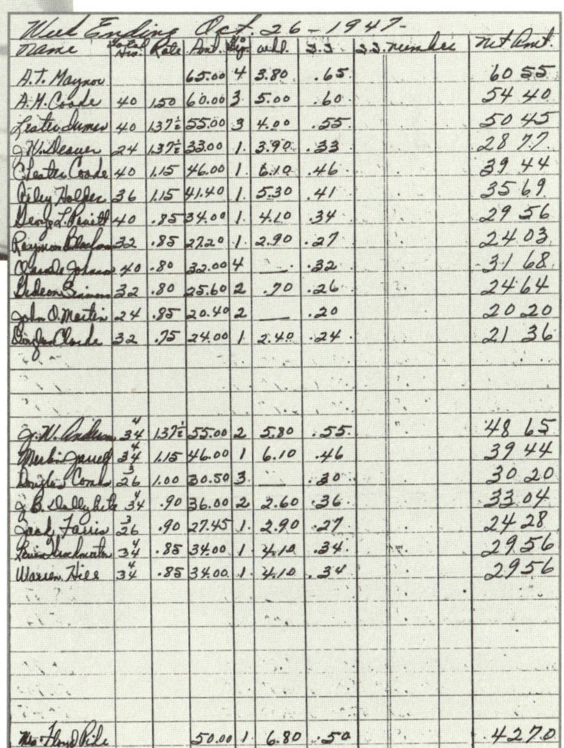

Week Ending Oct. 26 - 1947 -

| Name | Total Hrs. | Rate | Amt. | No. Exp. | W.H. | S.S. | S.S. number | Net Amt. |
|---|---|---|---|---|---|---|---|---|
| A.T. Maynor | | | 65.00 | 4 | 3.80 | .65 | | 60 55 |
| A.H. Coode | 40 | 1.50 | 60.00 | 3 | 5.00 | .60 | | 54 40 |
| Lester Dumer | 40 | 1.37½ | 55.00 | 3 | 4.00 | .55 | | 50 45 |
| J.W. Deaver | 24 | 1.37½ | 33.00 | 1 | 3.90 | .33 | | 28 77 |
| Chester Coode | 40 | 1.15 | 46.00 | 1 | 6.10 | .46 | | 39 44 |
| Riley Holder | 36 | 1.15 | 41.40 | 1 | 5.30 | .41 | | 35 69 |
| George L. Pruitt | 40 | .85 | 34.00 | 1 | 4.10 | .34 | | 29 56 |
| Raymon [illegible] | 32 | .85 | 27.20 | 1 | 2.90 | .27 | | 24 03 |
| Claude Johnson | 40 | .80 | 32.00 | 4 | — | .32 | | 31 68 |
| Gideon Simon | 32 | .80 | 25.60 | 2 | .70 | .26 | | 24 64 |
| John O. Martin | 24 | .85 | 20.40 | 2 | — | .20 | | 20 20 |
| [illegible] Clarke | 32 | .75 | 24.00 | 1 | 2.40 | .24 | | 21 36 |
| J.W. Anderson | [4] 34 | 1.37½ | 55.00 | 2 | 5.80 | .55 | | 48 65 |
| Merlin Jarrell | [4] 34 | 1.15 | 46.00 | 1 | 6.10 | .46 | | 39 44 |
| Douglas Combs | [3] 26 | 1.00 | 30.50 | 3 | — | .30 | | 30 20 |
| J.B. [illegible] | [4] 34 | .90 | 36.00 | 2 | 2.60 | .36 | | 33 04 |
| Jack Farris | [3] 26 | .90 | 27.45 | 1 | 2.90 | .27 | | 24 28 |
| Louis Throckmorton | [4] 34 | .85 | 34.00 | 1 | 4.10 | .34 | | 29 56 |
| Warren Hill | [4] 34 | .85 | 34.00 | 1 | 4.10 | .34 | | 29 56 |
| Mrs. Floyd Pike | | | 50.00 | 1 | 6.80 | .50 | | 42 70 |

Opal Pike's early bookkeerping.

*I am a lineman for the county,*
*And I drive the main road;*
*Searchin' in the sun*
*for another overload.*
*I hear you singin' in the wires,*
*I can hear you thru the whine—*
*And the Wichita Lineman*
*Is still on the line....*

—*Jimmy Webb,*
*Wichita Lineman*[4]

# 10

# On the Job

During those first twelve months, predictive of the future, the numbers of crews and their locations began to increase. The crews then were large: generally, at least seven or eight men; sometimes, even ten or twelve, whereas, today, a crew may occasionally consist of one man and his truck. In addition to the work in Mount Airy, Pike moved first into Winston-Salem; then other cities and towns in North Carolina — Leaksville, now called Eden; Madison; Gastonia; Charlotte; and North Wilkesboro.

In *Duke Power Company: The First 75 Years*, Maynor explains:

> The maintenance program was more than line and pole replacements, although that constituted a great part of the overall objective. Just before the war started, Duke had begun to increase the voltage on its distribution facilities by replacing the antiquated Delta-connected system with a "Y" connected system.... Post-war maintenance was a perfect opportunity to complete that conversion in equipment.[5]

From the beginning, Pike's crews have always gone anywhere there was work to be done. Regardless of the distance, the company's mobility has been essential to its success. For the crews working in North Carolina's rural environment, accommodations early on were often found wanting. This was true for Roy Thomas who began working with Pike in March '47. "Six of us on the crew stayed at a boarding house in Spray, North Carolina, which adjoins Leaksville. The place had little steel, army-type cots set up in one room of the house. It cost us $7 per week, and that included breakfast, a packed lunch, and dinner!"

Roy Thomas's situation was a typical one; Tommy Jarrell's was not:

> We put on a crew in Gastonia, North Carolina to set and frame poles. James Anderson was the foreman, and I was one of the crew members. The first day there, we worked until really late. By the time we got to the hotel, there was only one room left. Mr. Pike asked the man at the desk if they had any rollaway beds; they did. But even after they filled up Mr. Pike's room with beds, we were still one short. Mr. Pike had *me* sleep in his bed with him that night! The next day ... we got those rooms!

While locating accommodations was a frequent difficulty, locating affordable equipment was a relentless challenge —

especially for a company operating on a shoestring. From those last three weeks of December,'45, to January,'50, the number of Pike employees increased from the original two to eighty-eight. Before 1950 ended, the employee roster had passed one hundred. "Floyd went anywhere possible to find equipment that had any potential to operate," said Emmett Forrest. "He and William Anderson went to Indianapolis in the late '40s, and bought some old line trucks from Cohen Used Equipment Dealers. They had already been retired and sold by Bell Telephone and were really troublesome! Always overheating. Floyd and William were several days getting them back here."

Zack Blackmon, Sr., Pike Electric vice-president, Central Division, retired, began working with the company in December, '47, when Floyd Pike hired him away from the Blackmon family store to help to replace a crew that would not work on a holiday. Blackmon recalled using those trucks:

> At that time, our first mechanic, Dillard Bullin, worked night and day to get those old trucks in shape, so we could get them out on line crews. They were a short wheel-base truck ... got around fairly well in rough places. They were short on power, too, and, of course, worn out by the time we got them. But they sufficed, and we started line crews. I had a line crew, one time, with seven men on it. We'd pack five fellows in the back, with all the materials — and every thing else that you've got up there. We looked like gypsies, moving, trying to get to the job!

With the difficulties in procurement and the tremendous financial investment that equipment has always required, it is understandable that, from the beginning, Floyd Pike was adamant that individuals should treat equipment with utmost care and respect — a policy that is standard in the company's operation today. One individual who really shared Floyd Pike's feeling was Shorty Clawson, who joined the company in '46. He taught, by example, how to care for equipment. Donald Anderson, supervisor, Central Division, verified Clawson' s reputation when he declared, "With Shorty, even the auger was clean! We had to clean it as soon as we finished digging so no dirt would fall into that truck!"

Clawson even had designated places on the back of his pickup for the crew members to sit for lunch, and they were required to sit on paper! "Rarely was anyone else even allowed to drive Shorty's truck," Donald Anderson added.

Tommy Jarrell said that individual responsibility was explained early on to him by Floyd Pike:

> Mr. Pike got pretty upset with me one time. I had just become a foreman and had started a crew providing service to Duke Power out of Rutherfordton, North Carolina. The job site was actually at Lake Lure, in the North Carolina mountains. We finished the day and were on our way in. Coming down the last part of the mountain, the line truck's brakes — just worked on over the weekend — gave out again. It ran away with my driver, turned over, and rolled back up on its wheels; tore up, too bad to use! While a patrolman was inspecting that line truck's damage, he noticed that a headlight was out on my pickup. He told Duke Power's personnel there that he didn't believe that the maintenance was kept up on the Pike trucks. That word got back to Mr. Pike, and he wanted to talk to me. When he asked me why my pickup light was out, I was kind of aggravated and said, "I don't know, Mr. Pike. It might have been out when you sent me the truck!" He got very upset with me then. "If I send you something out there that doesn't even have any wheels on it," Mr. Pike said, " I expect you to get it running right!"

The demands on equipment were secondary to the extraordinarily difficult working conditions of the crews. "All holes were dug, poles were set, and wire was pulled, literally, by hand," said Tommy Jarrell. "There were no buckets," Zack Blackmon, Sr., explained. "Bucket trucks had not yet been developed. Linemen climbed – sometimes as many as seventy-five poles a day." Charlie Pack, safety supervisor and long-time foreman in Elkin, North Carolina, spoke about the difficulties from experience, when he talked about working on a line crew under Pike foreman Wade Martin: "He'd set you off back in those days, and you went to work. When the man turned you loose, you dug holes until dinner time, and then he'd come by and pick you up!"

Typically, the duties of a large crew were divided so that two or three men were cutting right-of-way while two or three others were digging holes for poles; and, at the same time, other men were framing poles. Depending on the size of the pole, from four to six men were required for pulling the poles into place. For poles larger than 30 to 35-feet, additional men were required. However, usually five men could easily set smaller distribution poles by piking them in.

A "pike" is a long pole, fifteen to eighteen-feet long, with a sharp point in one end that will pierce the pole being set. After a pole is raised by hand, pikes can be used to hold the pole vertically, while it is guided over the hole.

When setting poles by hand, crews usually used three or four pikes, along with a tool called a "dead man." A "dead man" is a rod with a steel crescent attached to the end to support a pole while it is being set. Opal Pike said that she once received a call in her office from a shipping company that was stymied by a package with its contents labeled "dead man." Chuckling, she said that she instructed the caller, "Please deliver. I have ordered one!"

Using devices such as these, crews worked and disturbed very little along the way. "This has always been our method," Roy Thomas said. "We rarely ever took a truck out across a planted field or any other property if we could avoid it. Most of the time, we'd pike those polls in, to keep the trucks off that farmer's crops!"

Undeniably, the work was made somewhat easier when additional equipment could be used. Roy Thomas added that whenever the crews could get trucks to the sites, "We would gladly set those poles with a line truck, a winch-cable, and an old, conventional, three-legged boom."

Some pieces of equipment used today are so fundamental to the job that it is hard to imagine working without them. Chain saws, for instance, replaced cross-cut saws and axes. Thomas said that the first chain saws were really crude when compared to those used today:

> The first chain saw that I used on a job had two handles – like a bicycle, attached to a motor. Two men were required to operate it. The blade would rotate from horizontal to vertical. You'd first flop the blade over and cut the tree off with the horizontal. Then after the tree was on the ground, the blade was turned 90°, and you'd be cutting straight. The thing was really difficult to start. But when it got cranked and running, it would do some sawing!

It is also very hard to imagine field work without the use of air-compressors. However, before their development, crews dug holes in rock with a hammer, a piece of steel, and a generous amount of dynamite. A crewman would hammer a steel rod until it would go no further. Dynamite was then used to make the digging easier. Roy Thomas said that the crews simply repeated the process until the hole reached the required depth. "Usually a rock hole for a 35' pole needed to be about 5' deep. Sometimes, if it was really bad, they'd let us get away with 4½-feet!"

The men on the crews of the fledgling company worked — and worked extraordinarily well — in their difficult conditions. And they were learning on-the-job to work as teams. "Floyd and 'Horse' were always coming to Winston, visiting the crews down there," said Roy Thomas. "Everybody in the crew was always glad to see them. They were both very helpful, and they would do anything they could for you — personal or otherwise — just an ideal working situation."

No matter how exciting it was to be a part of a new endeavor that was expanding by the month, the jobs themselves demanded skills of men who, for the most part, were still in the process of "becoming" experienced. The affinity between line work and danger is constant. The reality of the danger struck an unforgettable blow to the company in late August,'48 when Pike suffered its first fatality. The accident happened in Leaksville, where Berkley Snyder's crew was reconductoring. T.D. Fulp was a nineteen-year-old groundman who accidentally sliced through the insulation as he cut a wire. "When the foreman saw what was happening," Roy Thomas said, "he ran up and kicked the bolt cutters out of the boy's hand. But it was too late." He said the crew members did everything they knew. "We worked on him with the old 'arm-back,' pressure-resuscitation method that we used at that time, and we were only about a few blocks from the hospital. It was an especially sad day for us individually and as a company," said Thomas.

"This was really a tough time for all of us," said Opal Pike. "Floyd and I had never expected anything like that to happen. We wondered if it was even worth it to go on."

*We were making ends meet, but it was hard.*
*We didn't have any money.*
*We couldn't invoice and wait a week to get the pay.*
*We had to get paid when we finished the job.*
*When we ended up the work week every Friday,*
*we would go to Charlotte.*
*Floyd would give Duke Power the invoice,*
*and they'd give him a check.*
*We'd come straight back*
*and put the money in the bank in Mount Airy,*
*so our checks would be good.*
*We turned our accounts-receivable over pretty fast!*
*—Opal Pike*

# 11

# Making Ends Meet

Handwritten in Opal Pike's ledgers of the late'40s, the list of employees lengthened. These well-remembered names remain important to the company today:

Jack Farris; William Anderson; Lewis Throckmorton; Ralph Tucker; George Pruitt; Gideon Simmons; Raymond Blackmon; Douglas Combs; Warren Hill; the Lankfords: Leonard and Haywood; Emmett Tucker; Lowell Anderson; Bishop Key; the Griffins: J.B, and W.S.; Cecil Calhoun; Claude Chilton; the Coles: S.J. and L. D.; J. B. Clawson; Roy Graybeal; William Mitchell; J.D. Shumate; Mack McCall; T. D. Fulp; the Mears, Ray and Grady; Lindsay Beck; John Reddic; Arlie Holder; the Coes; John Martin; Eugene Jessup; Rufus Niston; Fred Caudle; Cecil Fletcher; W.H. Clement; Ralph Hudson; W.E. Harmon, and many others …

At times for the young company and its employees, a paycheck, even with supplement, must have seemed tenuous. Glenn Pike, retired vice-president, Commercial Wiring Division, joined the company in '49 at the age of seventeen. "My foreman was Howard Farmer when I went to work in Leaksville. Floyd drove me over there in his '40 Plymouth coupe. The crew size was around eight or ten people. The starting pay was 85¢ per hour, less room and board. Top lineman's pay was $1.35; and I thought if I could someday make $1.35 an hour and be a top lineman, I would have it made financially!"

James Anderson said that early on there were often long periods between raises. "During one of those times, Floyd came up and said he was going to Charlotte and get me a raise." Anderson said that when Floyd Pike returned, he talked about several other things for a while before he finally told him to look in the trunk of the car. What James Anderson found, was not what he was expecting. "James, I couldn't get you a raise," Floyd Pike explained, "but I did bring you an anchor-bore!" "Well, it was almost worth as much!" laughed Anderson.

To aid with the cash flow, Floyd Pike bought a couple of car-haulers and eventually added two others. The company got paid for hauling so many pounds of cars per mile. A.L. McAlexander, retired assistant vice-president, pinpointed the dates of this segment of the company's history — and his own — to the moment:

> I always thought that one of the reasons Floyd needed me as a mechanic was to keep up those road tractors. I had worked out my notice to Fowler Motor Company, an automobile dealership in Mount Airy. When I quit for the week on Saturday noon, I loaded my tool box in the back of my car, and the lid wouldn't close. I decided, rather than to struggle with unloading

> the tool box, and having to load it again, I would just drive on to the Pike office. Mr. Pike was there, and said he needed me that afternoon, if I wanted to go to work. I went to work on Saturday afternoon, March 11, 1950, at 1:00 p.m.

The car-haulers would usually run a short run and a long run. Roy Thomas and William Anderson ran the route dispatched from Crawford Transportation in Ironton, Ohio, to Detroit to get loads of Chrysler cars. The men delivered cars to all of the Chrysler Corporation's dealerships in West Virginia, North and South Carolina, and sometimes, Kentucky. "Back then, Chrysler was a pretty doggone good car," said Roy Thomas. "There was a dealer in about every town of any size, so it kept us running and helped out with extra cash for a good little while." Coincidentally, those trucks began to age about the same time the Chrysler Corporation experienced a strike. By '52, William Anderson and Roy Thomas returned to line work, and the trucks were sold.

## PART-TIME HELP

To help make ends meet in the early years, part-time help was hired in the summer months. This was a beneficial situation for both employer and the part-time employees. "The first time I came to work was in the summer of 1949," Claude Brinkley recalled. "I spent that summer cutting right-of-way in Leaksville." He said that the next three summers, and the summer of '55 were his "summer training periods." Brinkley said that he really didn't remember asking for a summer job, "My Dad and Mr. Pike, and, I'm sure, Mr. Maynor had something to do with it. Because each summer, the first day out of school was my first day working on a line crew or with the right-of-way crews."

The schedule was the same for other men whose careers later followed paths far removed from power-line construction. While Fred Folger was in law school, he dug holes for "Horse" Maynor in the summers. Ernie Randleman, whose father ran the grocery store in Mount Airy where young Floyd Pike sold his rabbits, became a successful pharmacist and businessman. Ernie Randleman's first job was with Pike.

Part-time helpers were treated no differently from other employees. When their summer's work ended, those young men understood the definition and procedures of hard work, an understanding attained through experience, but not without guidance. For some workers, the lessons learned from their part-time employment extended into life-long significance. In the following reminiscence, Dr. Paul D. Webster III, originally from Mount Airy, and now medical professor at the Medical College of Georgia, Augusta, discussed how important those summers spent working for Pike were to him personally:

*Dr. Paul Webster, Medical College of Georgia. His summer job was with Pike.*

> With great pleasure and satisfaction, I reflect on my contacts and memories of Floyd S. Pike. My contacts with Floyd were unique for several reasons.
>
> I first met Floyd when he was a lineman for Duke Power Company, probably in 1940-41. Duke Power replaced a utility pole in our front yard when my family lived on Banks Street. Several of us watched the linemen replace the pole, redo the wires, etc. The neighborhood kids were awed by the whole process. One of those linemen was a young kid who did not look a great deal older than we. This man was Floyd S. Pike — a man whom I would get to know much

better in the future, a man who would serve as mentor and play a role in my own development, which would be far greater than expected from the time of contact. The supervisor, or crew foreman, was, I believe, Floyd's brother, Jim [James] Pike.

The next time I remember meeting Floyd was when I learned about the opportunity to work for him during the summer months. I was a high school student and wanted an outdoor job for physical conditioning so I could better participate in football the following fall. I played football in high school, and received a football scholarship to the University of Richmond. However, I stopped playing after one year because of injuries sustained to my knee, and the fact that I weighed only 165 pounds, and the average lineman in the Atlantic Coast Conference must have weighed 200 pounds. Freddie Folger, a football notable from Mount Airy High and Duke University, and a number of other young men wanted outdoor jobs, found that opportunity with Pike Electric Company. I worked for three years for Pike: 1947, 1948, and 1949, and then started college, medical school, and years of residency training eventuating in a career in academic medicine.

At that time, Pike Electric was much smaller, and we could expect to receive an occasional visit at the work site from Floyd, who would either drive up in a gray 1942 or 1946 Plymouth, or just walk up wherever we were working. Holes for utility poles were dug by hand in those days. At times, it was an easy job, if you had soft soil, or nearly impossible job if you had "a rocky hole."

I remember several conversations with Floyd and will recount these. Most important is what I carried away from the conversation. As I think back, each time we had such a session, it was for me a mentoring session, a session that taught me a significant lesson about the work world and how to succeed.

One day Floyd stopped his car near the road where I was working, got out, and looked into the hole that I was digging. After a few weeks on the job, I had become quite facile, and could dig a very small hole fashioned to fit the size of the pole. On this occasion, I dug such a small hole that Floyd said the pole wouldn't fit into the hole; and if it did, you could not tamp dirt around the pole. I was sure that hole and pole matched, even though there was little space. Floyd was pleasant, but could get his point across very quickly. He had a way of being positive, and focused. Needless to say, the hole was immediately enlarged to fit his specifications. We chatted some more and he left.

After I had been on the job for a month, my second conversation with Floyd transpired. I had inquired from co-workers what their hourly wage might be and found that I was paid 5¢ less than my peers. I knew that I was digging as many holes, and doing as much work as they, and, in fact, outdid a few. So after about a month on the job, I asked the foreman "Horse" for a raise. Maynor told me, if I remember correctly, that it was a little early and premature for such a novice to receive a raise. My comment was that I was doing the same amount of work as the others and expressed a feeling that I desired equal compensation. I guess this was my first expression of the concept of equal work with equal pay. "Horse" indicated that he would talk to Floyd and let me know the results.

A few days later, Floyd came to the work site in his gray Plymouth; he called me aside, and told me that "Horse" had discussed with him my need for a raise. Floyd was smiling and seemed to be amused by the whole affair. He stated that he would grant me the 5¢; but, most importantly, he pointed out that I had looked out after myself and presented my own interest. He commented that it was important to learn to look after one's self because "no one else would." Know what you are worth, know what your output is, and be willing to press for adequate compensation or recognition. This was a very valuable lesson for a sixteen-year-old boy. At that time for me to press my interest as a sixteen-year-old to the foreman of the line crew might have been interpreted as brashness. On many occasions thereafter, Floyd would joke with me about "the raise."

Later in the summer, Floyd visited the worksite on a hot August day late in the afternoon. I had already nearly completed the hole that I was working on, and I believe I was relaxing or sitting on a log! Floyd walked up and pointed out that when the foreman or the boss comes around, you should look busy – get in a strain, even if you have to pick up something to do it! It will give a good impression. I learned this lesson well, and have used it many times. For example, earlier in my career when I was directing a very active research laboratory, we were scheduled to have a visit from a team from the National Institutes of Health to determine future funding. The day the visitors from the NIH came, we had "the busiest laboratory they had ever seen". We "impressed the boss;" and we received our funding. This lesson learned from Floyd has served me well many times, over many years.

During the second year, I spent part of the time cutting right-of-way. Floyd visited the work site, which I believe was somewhere in the mountains of North Carolina — very rough area, where we were clearing right-of-way for a new line. I must have made some comment during Floyd's visit about the lack of challenge involved in the work, and Floyd responded with the following lesson. He indicated that the reason he was successful when he started contracting for Duke Power was that he learned and understood the job requirements one or two levels above his own. He stressed the importance of understanding and being able to perform your supervisor's job, of even his supervisor! Thus, when it came time for a promotion, or to establish an independent business, you would be successful.

This advice seemed very sound to me and I have exploited that lesson on many occasions. I have tried to understand the task and performance of the next levels of responsibility. In an hierarchical system, and if an individual desires to excel, the ability to know and understand (and even perform) the next level of tasks is a vital one for the ambitious person who wants to get ahead. I learned this lesson well from an expert — an individual who used these attitudes and philosophies to develop a very successful business.

Another lesson I learned from Floyd centered around the work site. I projected my work schedule for the day and was pacing myself appropriately. Floyd came up, said hello, and we discussed activities of the day. I guess he asked me how many holes I thought I would dig during the day; and I looked at the soil and terrain, and stated that I could dig eight holes on that particular day. Floyd volunteered that if he had eight holes to dig in a day, he would try to dig them all earlier in the day so that he would have time at the end of the day to accept other tasks or opportunities as they might present themselves. He encouraged me to organize the day's work and to get the job done as early as possible so that I could move on to bigger or better opportunities. I have always tried to get as much of the job completed as early as possible so that other opportunities could be exploited. I have shared this philosophy with house staff and medical students, encouraging them to eagerly accomplish today's tasks so they would be free for the next opportunity.

Perform the job according to specification; know what you are worth and look after your own interest; understand your job and the requirements of the next two levels of supervision; complete your tasks as early and as eagerly as possible, so that you will be free for the next opportunity; project industry when the boss visits — that may be the only visit and impression he will carry away for the day, week, or month.

When I started working for Pike Electric, there were only several crews. One crew used a Dodge truck. The first mechanical hole-digger was Army surplus. Needless to say, holes were dug much faster with this machine than by hand; but I wonder if those doing the labor had the opportunity to learn as much. The enterprise grew; and the demands on Floyd's time increased. Future workers were probably unable to develop this close relationship with "the boss." I was indeed fortunate, in that I gained much more than I was hired to do. Many years later, after I had finished medical school and was a professor of medicine, I had an opportunity to review my experiences with Floyd. I took great pleasure in telling him of the lessons I had learned. I expressed my deep appreciation and indebtedness for his mentorship. He was pleased.

*The company's growth, overall, I'm sure,*
*far exceeded what any one of us*
*would have thought in the early '50s.*
*But under the direction of Floyd, as time went on,*
*we began to believe – and we found out–*
*that we could do just about anything.*
*That's had a lot to do with company's success.*
*—Glenn Pike*

12

# Incorporation

Increases in the numbers of crews, customers, and locations of work were accompanied by an escalation in the amount and the complexity of the company's bookkeeping. Emmett Forrest joined the company in 1949 to assist with office management and the company's accounts. "That was one of the greatest days of my life," Opal Pike said. "By then, my job had become overwhelming. Emmett had been helping me at night and on weekends, and then he came in and just took over almost everything."

Emmett Forrest was Opal Pike's brother and Floyd Pike's closest friend, their family confidant, and he served as the company's chief financial officer throughout his career. Floyd and Opal Pike, together with Emmett Forrest, set plans into motion that enabled the company to receive its charter and certificate of incorporation on July 1, 1953. Forrest was then appointed assistant secretary and treasurer.

Emmett Forrest designed the familiar and fitting logo that remains synonymous with the company today — a power line, bisected by a lightning bolt, set within a circular seal. The words, "Floyd S. Pike Electrical Contractor, Inc., Mount Airy, NC " are inscribed around the circle's inner edge.

Floyd and Opal Pike and Emmett Forrest were the only officers within the small corporation. They were the first three stockholders, and they comprised the company's first board of directors.

The company's minutes indicate that at the time of incorporation contracts were held not only with Duke Power Company but also with the Appalachian Power Company and the City of Danville, Virginia.

Dedication and confidence within the company were growing. By the company's second stockholders' meeting, one year after incorporation, the original three stockholders had been joined by twenty-one others. Pike's incorporation marked a point in its maturity and carried out an essential belief of its founders. Floyd and Opal Pike were determined that all of the individuals, who had shared the hard work and contributed to the company's success thus far, would also share the financial rewards.

Given the evidence of the working relationships that had developed within the company, employee-stock ownership was a natural progression. The participation of employees as stockholders in the small, growing business demonstrated both their courage and their confidence in management's capability to obtain the work and steer the course, and in the field personnel's competence to get the job done.

Together, all of the original stockholders shared a mutual objective: the success of the company and its people as a whole. Individually, those shareholders substantiated their sincerity with a financial commitment. While such a commitment certainly lessened their take-home pay at a time none of them could easily afford it, the resolve of those stockholders — expressed and built — their expectations of the future. The commitment that they made then was the foundation of the company that continues here today. Employee-stock ownership has been advantageous to all concerned.

"The big benefit in making stock available to all employees who had had a certain amount of service was in financing it," said Haywood Merritt, Jr., retired executive vice-president. "A certain amount of an employee's pay was taken out over the course of a year, and the people who took advantage of that offer — and that was most of them, I think — saw the value of their stock just going up, almost geometrically." He said that the stock's value "multiplied and multiplied and multiplied" and became quite valuable. "This created a strong pool of people who were very, very loyal to the Pike Company."

*Floyd Pike and Emmett Forrest, Jr. attended many meetings of the Public Utilities of the Virginias at the Greenbrier, White Sulpher Springs, WVA.*

*Well, the first memory I ever had of Floyd*
*was when he was driving into the*
*garage at the old Dodge place.*
*He came in, introduced himself, and we talked awhile.*
*At that time,we were servicing all of his trucks,*
*including the one or two used ones,*
*that he had bought from us.*
*All the Pike trucks were several years old at that time.*
*One day he came in and asked,*
*"How 'bout selling me one of those new ones?"*
*He had always paid in full for the used ones.*
*With the new ones, Floyd needed financing*
*from the time he bought the truck,*
*— until he got it equipped and ready to go on the road —*
*until he could get the whole package financed.*
*That was the way we started.*
*I sold him trucks for thirty years after that,*
*until Dodge stopped making a truck*
*that was big and strong enough to pull, with a line-truck body on it.*
*—F. D. Forrester*

13

# Keeping Them Running

In the beginning, necessity determined that the best equipment Floyd Pike could buy was whatever equipment he could find and afford. As rapidly as financially feasible, and as quickly as equipment for line work was developed, Floyd Pike established a precedent which demanded that only the best materiel available would be bought or made for his company. Providing Pike's personnel with the best and safest equipment is integrally connected to the company's record for service.

Equipment and its repair has always been a responsibility of the company's garage personnel. During the company's first few years, any money available was used to purchase the trucks and equip them, and there was usually little money left over for acquiring the tools necessary to keep the trucks running. The adaptability and ingenuity required and demonstrated by Pike's rudimentary garage operation during the time of the company's early growth is exemplary.

A.L. McAlexander's initial responsibilities of "keeping car-haulers on the road to help with the cash flow" quickly changed into the effort of establishing a garage operation capable of providing "anything at anytime" that a crew might require. Working in the first office-garage, McAlexander said the space was so small that he did not have room to work on some of the larger trucks if the door was closed. His assistants, Mondale Payne and Edgar Pike, Floyd Pike's oldest brother, stored supplies and repaired tools in a barn in back of the building. Sometimes when Ray Coleman was not helping Opal Pike with the books, he also helped with supplies.

*Edgar Pike, oldest brother of Floyd Pike.*

Although the tool and supply room was actually a barn, it was probably adequate at that time. A.L. McAlexander said the total garage equipment inventory consisted or no more than "a couple of vises, a borrowed air compressor, a transformer-type welder, and a coal stove for heating." Later on, Floyd Pike told him to buy a badly needed floor jack. "Then I

*This tiny space was the office garage.*

could get a pickup truck on the floor jack, slide the truck around to the right, drive a line truck in beside it, and I could have work going on two vehicles at the same time!"

A real difficulty with the office and garage occupying the same building occurred after the company settled on yellow as the vehicle color. Efforts to confine yellow spray-paint — just to the garage — were not always successful. McAlexander also noted that an additional worry was the fire hazard that accompanies ordinary garage work — with welding and burning, and,of course, the painting.

When more space became necessary for both the office and the garage, Floyd Pike instructed Joe Fulk, McAlexander's assistant mechanic, to grade an area of adjoining land. Pike foreman, Tom Scott, came off a crew and built a small garage with a tool and supply department and a separate room for painting. McAlexander explained that the new facility not only gave them "4200 square feet of elbow room," but the separate painting room encouraged garage personnel "to paint everything yellow — whether it needed it or not!" And, the new work space was heated, after he repaired an old furnace. The new building had many conveniences that included a bathroom and an enlarged phone booth with a writing shelf that served as an office. "With that feature, one could actually hear on the telephone at the same time work was going on!"

Along with the additional work space, the company even budgeted $2,000 for McAlexander to buy some much-needed equipment. The air-compressor, the motor-driven DC welder, the hydraulic press, and two chain hoists that he purchased then are all still in use.

A few years later, McAlexander drew up plans for a much larger, well-heated garage that doubled the size of the operation. It also provided garage management with office and supply space. This facility stands directly across from the main office and is equipped with modern ventilation, plus filtering systems for painting.

From the beginning, service to the trucks at the job site or at a break down has been a fundamental duty of the garage personnel. Since McAlexander was "the garage" in the beginning, when a breakdown occurred, he "simply locked the doors" and drove to wherever he was needed. As more trucks were added to the fleet, he was frequently needed in more places. He made field calls driving his own family car, with the company paying him 7¢ per mile. "This went on," he said, "until the automobile began to smell of grease!"

McAlexander could not be everywhere at once. In 1960, a mechanic's truck was bought, and Joe Fulk became the company's first full-time field mechanic.

By the late '60s and early '70s, the volume of the company's transmission work required one field mechanic or more on each job. A.L. McAlexander explained that transmission work is so hard on equipment that "a field mechanic usually puts in more hours on a job than anyone else — other than perhaps the supervisor and top management."

## PURCHASING ADDITIONAL EQUIPMENT

Only months after the company began, Pike crews were sent to work in North Wilkesboro, North Carolina, where Floyd Pike met F.D. Forrester, for many years, the owner and operator of Wilkes Dodge. The acquaintance grew into a friendship warmed by trust and mutual success. Opal Pike described Forrester as one of those people who had been very, very helpful along the way. "After Floyd unsuccessfully tried to buy a new truck 'on-time' from the local dealer in Mount Airy, Forrester's help with the financing allowed us to buy the trucks we needed," she explained. "He would let us pay him, just like we would pay the bank!" Opal Pike said that, later on, they didn't have to borrow money to pay for those trucks. "We just paid him direct!" The reluctant dealer in Mount Airy often expressed regrets for his lack of trust in their ability to pay for that one truck — not to mention his loss in sales. Purchasing the trucks, however, was only the beginning.

When bucket trucks came onto the market, Pike was waiting for them. Everyone could see their potential for expediting line work. To actually work in one, however, required some getting used to. Charlie Pack declared that he would never forget his first experience in a bucket truck:

> Roy Thomas pulled that truck out near the shed where we kept the tools, leveled it up, and told me to get into that half-moon shaped bucket. He swung the thing around, and there we were, looking down, with the outrigger raised up. I wouldn't stand up towards the front – thought the thing was going to turn over with us. I was scared to death, I said, "Boys, I'll never work in one of these things!"

Working from an aerial device was indeed quite a change. John Brewington, vice-president, Fleet Operations, explained that the '40s and the '50s brought little change to the way power-line construction and maintenance were performed. "Large crews and rugged operations requiring lots of people-power were the norm." "Mechanization of this work began in earnest in the late '50s and '60s with the production of aerial equipment that had fiberglass buckets and booms to help insulate the operator from ground sources." The changes that occurred then were important ones. "A-frames equipped with mechanical diggers and the hydraulic-rotation digger derrick used today are still very similar to those early units," Brewington pointed out. The first buckets did not have hydraulic controls — they were operated with wheels and fiberglass rods.

"The great change in power-line equipement that John Brewington referred to," said A. L. McAlexander, "was simply an evolution." He explained that a line truck evolved from an ordinary truck with an ordinary enclosed body, but with some important modifications. Tool boxes were added on the sides; and a winch was mounted to the back of the cab, along with a three-legged derrick that could be set up to hoist poles and transformers. "Those changes did not occur without some mistakes, red faces, and expenses on the part of the manufacturers and, to a small degree, some of the users," said McAlexander.

There were slight improvements in the hydraulic digger after it was manufactured with a fold-over boom. But in the line trucks that the company used, the first big improvement came when the Pittman "Polecat" was developed. Linemen finally had a hydraulic rotating digger-derrick. Progressing from the "polecat's" 180-degree rotation, The Baker Equipment Engineering Company supplied Pike's fleet with continuous, hydraulic-rotation digger derricks as soon as the Telelect Corporation developed them.

Developments in the trucks themselves, in their mountings, and other modifications did not always parallel, and often lagged behind, the company's needs. Consequently, Pike has dealt directly and productively with design personnel and equipment manufacturers as progress has occurred. "In the evolution of electrical construction equipment," said Brewington, "the knowledge and input from actual users played an important role in the equipment and tools produced for our industry."

This fact has been demonstrated time and again by Pike in conjunction with the manufacturers of line construction equipment. For example, Baker Engineering and Equipment Company president Glenn Baker — and his father before him — worked with Pike in customizing truck modification and design for decades.

A.L. McAlexander confirmed that from the company's earliest days, "Floyd S. Pike Electrical Contractor, Inc., became somewhat of a proving ground for some of the new equipment." Pike often carried out test procedures for its customers by using new or experimental equipment on the jobs. "The results of these experiments were beneficial to everyone involved — the manufacturer, our customers, as well as ourselves," said McAlexander. "The findings led to improved products and improvements in the way we worked."

When Telelect's [now Simon-Telelect] "Captain" boom series was added to the fleet, company personnel worked directly with its designer to modify the boom to fit the particular needs of Pike crews. Floyd Pike, A.L. McAlexander, and Zack

Blackmon, Sr., made many trips to the factory to expedite necessary changes. McAlexander's input was helpful in determining how to mount the digger on the boom in a way that it would remain maneuverable, and not self-destruct, while digging in hard earth. He did a great deal of work early on to improve the performance of hydraulic systems, particularly in the smooth operation of truck radial outriggers.

In other modifications that have been made to truck-booms, a "claw-attachment" is a refinement that was suggested by Pike. This attachment allowed a crew to plumb a pole mechanically, eliminating "piking" that had always been performed manually. Zack Blackmon, Sr., recalled that the Telelect Company colloquially designated one of the claw arrangements, the "Zack Blackmon Pole Claw," because Pike, and later, many others, requested it so frequently.

As the transmission work increased in challenge and in volume, the company needed more equipment — tremendously heavy, specialized equipment — with added components that were required for stringing and tensioning electrical conductors. The early jobs often required A.L. McAlexander to devise some really ingenious mechanisms. "In some of the company's first transmission work, we were faced with the problem of removing copper conductor from a steel tower line and replacing it with new and larger aluminum conductor," McAlexander explained. "What we needed was a machine that would simultaneously pull and wind the old conductor wire on wood reels, while pulling in the new wire." He knew the job could be done by joining the old and new wires together and pulling them over rollers attached to the insulators, and he had an idea for expediting the task.

Two old gear boxes, a new power takeoff, six new sprockets with some new sprocket chain, along with a few fabricated, and some machined, parts were assembled in the proper order on an old surplus Marine Corps Ford truck, called the wire winder. "This machine operated just as anticipated — down to the over-heating of one of the gear boxes on long pulls," A.L. McAlexander explained. "The solution for the heating problem was not complicated. For a cooling effect, we simply poured cold water on it!" Later in the sequence of equipment development, Pike's use of the (20,000 lb. rated) Commander II for transmission work, provided a crew with a truck which not only had a rotating digger-derrick, but the Commander II had the capability to perform as a crane, a feature that served the company well, and often, on some very big and tough jobs of the '70s.

*I guess it was around 1954 or'55,*
*Floyd came up one day and asked me*
*to be a supervisor to look after all the work*
*that Haywood Merritt had going around Bluefield,*
*Abingdon, and Pulaski, Virginia.*
*When I said "OK.," he then asked,*
*"Who do you think would be the best foreman for your crew?"*
*I told him that I thought Fred Anderson, now our retired*
*supervisor, would do a good job. Floyd then talked to Fred,*
*but Fred didn't want to take the crew.*
*Floyd kept talking to him, and finally made him an offer:*
*"Fred, you take it for two weeks,*
*and I'll be back up here to talk to you.*
*If you don't want it then,*
*I'll get somebody else to take it."*
*Fred ran that crew for two weeks.*
*And, sure enough, when Floyd returned,*
*Fred took that job; and he held it for a long time*
*— about forty years!*

*—Roy Thomas*

# 14

# Real Expansion Occurs

A competent, maturing management team was doing the job and steering an ever-widening course. In the '50s and '60s, the work obtained for those in the field was increasing by leaps and bounds. Pike crews were working for the Carolina Power and Light Company and for Potomac Edison. Lawrence Allen, former assistant vice-president, had most of the streets torn up in Greenville, South Carolina, for Duke Power's underground conversion. Pike had obtained work with Monongahela Power Company, and the company had a contract with TVA. Zack Blackmon, Sr., had moved with his crews into Maryland to work for Baltimore Gas and Electric. "I took seven crews there in April, 1956, and the work gave out in September, '57. However in the following spring, a tremendous wet-snowstorm put the BG&E system 98 percent out of power," said Blackmon. "I went right back to Baltimore!"

As a result of the storm work, the company established a good working relationship that continued with BG&E for many years. "We had some crews up there through the '60s," Blackmon recalled. "We pulled out sometime in the early '70s during the big cutbacks of distribution line work, then we went back into Baltimore in the late '80s."

*Baltimore Ice Storm.*
*(Photo Courtesy: Richard & Glenda Simmons)*

Zack Blackmon, Jr., vice-president, human resources, recalled that his dad was spending so much time directing the work in Baltimore, that "we just moved there!" And so did many other families. "Bud Smith and his family; Howard Farmer and his; Blaine Bard; Charles Combs; and there were lots more," said the younger Blackmon. "One of the best parts of living there was the fun that Pikes' families had when we got together on the weekends and holidays."

Meanwhile, the company began working practically in its own back yard for the Surry-Yadkin Electric Membership Corporation, located in Dobson, North Carolina, twelve miles from Mount Airy. Kelly Hutchens, general manager, retired, said that his co-op hired a Pike crew in October, '62. "The men in the crew were: B. Hathcock; C. Johnson; B. Koone; D. Angel; J. Hicks; and Wade Martin, foreman. Hutchens added that the crew's rolling equipment was a conventional line truck of that era and a pickup truck, and that the "crew's field supervisor and our contact person was Zack Blackmon, Sr."

The nucleus of Pike's first generation of management and

supervisory personnel were men who worked their way up through the ranks, taking on more and more responsibility. One of those men was Haywood Merritt, a young electrical engineer, who joined the company only six months before its incorporation. He was in charge of a great deal of the company's expansion during this time. "When I was finishing up two years of service in the Army," Haywood Merritt said, "my father suggested that I talk to Floyd Pike who had become his good friend while I had been away." Although Merritt had worked at the North Carolina Granite Quarry in Mount Airy before the service, his father thought that Pike could be a growing company, and that the younger Merritt might like its type of work. "I did interview with Mr. Pike, and I was hired to work for the company in January, 1953." Haywood Merritt's first assignment with Pike was a joint endeavor with Duke Power Company:

> The first job I had was in the area of Gastonia and Belmont, North Carolina. I was the assistant to a man employed by a competitor of ours. The two of us were designing and drawing an improvement to Duke Power's electrical facilities near Gastonia. We were converting a 2300-volt line with small wire to large conductor, 4160 volts. I did this for several months, until we finished the project.
>
> The gentleman who was coordinating this for the power company was a man named H. E. Carter, who was a very, very important person in the relationship that we had with Duke Power. Mr. Carter took me from that job and sent me to the lake where several contractors had a training camp. What Mr. Carter wanted me to do was to make drawings which became a part of

*Lineman Training School at High Rock Lake.*

the first construc-tion manual of detailed drawings and specifications for distribution work that Duke Power ever had.

*Kitchen/Classroom at Lineman Training School at High Rock Lake.*

The contractors' training camp where Haywood Merritt worked early on was located at High Rock Lake, near Salisbury, North Carolina. Its existence was the result of a collaborative effort with Duke Power and several other area electrical contractors. Employees from the various sponsoring companies served as instructors. Linemen went to the camp for refresher training, or to learn line work in preparation for employment. Generally, a training session lasted one week.

On the campus, the men slept in bunk-houses and ate all meals in the kitchen/classroom building. On the grounds, arrangements of "short" poles allowed the men to learn how to hook-up transformers, move conductors, install hardware, and any other tasks required in line work, with their feet still "on the ground." In other areas, poles were set for "climbing" experience.

Instruction for hot-stick work was also taught there — how to use the devices properly: how to care, clean, and repair them. The joint training school lasted a few years, and then the responsibility for training was returned to the individual companies.

For many of Pike's linemen, hot-stick training was an important part of the school's program. A hot stick was a rod device that varied with its attachments. Hot-sticks were actually made of highly varnished wood, and the trailers that transported them had to be heated for dehumidification. These devices allowed a lineman to work closely with an energized line or conductor while on a pole. With the hot-stick's extension, a lineman was insulated from the wire as he performed his job. Along with becoming skillful in working with the stick itself, a lineman had to learn about numerous attachments and how to use them in allying, untying, cutting, pulling, or other requirements involved in the line work. Zack Blackmon, Sr., said that the company "started pioneering with hot-sticks as a method of doing line work." He explained that immediately after World War II, most voltages were 2,300 Delta in cities, and "linemen worked that energized a bit" when they could not pull it off for some reason. "In rural areas, with 12,000 volt lines energized, hot-stick work became necessary," said Blackmon. "We had seventeen hot-stick trailers that I knew about, and we were using them all over our system. We got to be pretty expert at it!"

Although hot-stick work was dangerous, almost all of Pike's "A" linemen – linemen with seniority in skills and experience – were capable of using the devices. However, with improvements in rubber goods and the use of insulated buckets, linemen rapidly began to rely on gloves and bare-hand techniques instead of hot-sticks for higher-transmission voltage work. Hot-sticks continued to be used for transmission maintenance.

Incidentally, in the '80s, hot-sticks made a comeback when distribution voltages were increased into the 25 to 35 KV range and, where utilities had miles of lines in many non-rural areas that crossed private property and in congested, commercial areas that were inaccessible to equipment. Probably more a important reason for reviving hot-stick use was that line maintenance and construction could continue without interrupting electrical service. In the '80s, the company's newsletter, the *Pike Line*, proudly reported that the company still had linemen on the job who were trained, ready and able to do hot- stick work, having done it thirty years before. The company also responded to the method's reoccurrence by buying new trailers and by training additional personnel to handle the increase in the demand of this type of work. Some of the instruction was provided by Appalachian Power Company, for whom Pike began contract work in 1950.

## NEW ENDEAVORS IN THE NORTHERN DIVISION

The company's first contract work for Appalachian Power Company began in the Marion, Virginia, area. Soon, Pike obtained the contract to provide service to APCO's entire Pulaski, Virginia, division. And when the company was awarded a major job in Bluefield, West Virginia, — converting the north side of town from 4KV to 12 KV, — the company's work really expanded.

"The Appalachian Power contract was really a big step," according to Roy Thomas. "We only had two crews working in Bluefield when I got there. Lewis Throckmorton was foreman of one of those crews, and Odell Klore had the other one." The job involved extensive reconductoring and pole-changing, and Pike's crews rebuilt almost every foot of the existing lines in the area. "Just hundreds and hundreds of new poles and new wire went in."

Not long afterwards, crews were added in the Abingdon, Virginia, district and the Welch, West Virginia, area. Pike's work then covered the non-union divisions of Appalachian Power. The huge increase in the company's work really increased the responsibilities of Haywood Merritt. "Not only was I overseeing the work as it was being done, but I was also responsible for putting crews together, keeping track of the crews, billing for the work after it was done, and determining what kind of profit or loss we were making." He explained that all the work done for Appalachian was "unit-priced," which meant that the work had to be inventoried in the field, and the unit prices that had been negotiated, were used to calculate the invoices.

As the work for Appalachian Power neared completion, other work for Pike's crews had to be found. Floyd Pike transferred Roy Thomas from Virginia to Georgia to explore the possibility of working there.

## WORKING "DOWN SOUTH"

In the mid -'50s, Larry Graham, former Rural Electrical Membership Cooperative[EMC] manager in Cobb County, Georgia, called Floyd Pike to request the rebuilding of his entire EMC system. With that call, the Pike company had a "foot in the door" in Georgia, and crews were on the job five days later!

"At the time, I was told by forty other EMCs," Larry Graham said, "that the work was too dangerous and would injure many people." He said that he had confidence in Pike's people and that his decision to hire Pike had been a wise one — a fact long since recognized by many of those who had doubted him. "Since that time, all 41 EMCs have contracted with Pike," Larry Graham said. When the work in Cobb County necessitated additional help, Pike General foreman Bill Griggs went to Elberton, Georgia, and hired Glenn Dobson, retired Pike assistant vice-president. Colbert Beasley was also working in Georgia by the late '50s.

Good for all involved was the presence of Pike's personnel not only for the reconstruction, but also during storm trouble. Graham said that Pike crews quickly responded to his co-op's needs anytime they were called for storm restoration, and that they always worked with pride and without complaint. During a particularly bad ice storm in which Pike's personnel had worked for long hours without much rest, Graham recalled a Pike crewman's unforgettable question — the closest thing to a complaint that he ever heard: "Don't you wish we all lived in Miami, Florida?"

In the late '50s, Pike crews were also moved into the Deep South. Through Floyd Pike's participation and support of the Southeastern Electric Exchange, the company's reputation had become known in Louisiana to Gene Lancaster, who was an assistant vice-president with the Pike company at the time of his death in '72. At that time, however, Lancaster was working for the Louisiana Power and Light Company, and he hired Pike to string conductors from a river-front tower at New Orleans, across the Mississippi River, to a tower that stood on a plant-site of the Kaiser Aluminum Company.

Construction of the towers was almost completed when Joe Fulk and a crew of fifteen men arrived to begin the job. The astonishing part of this job were the tower sizes — on one side of the river, the tower was 480' high; and on the other, the tower stood 460'. Zack Blackmon, Sr., said that it really "separated the men from the boys to work at that height!"

The work schedule was the first problem. Blackmon said that the crews had to work with the Army Corps of Engineers who controlled the river. "They informed us as to when there would be no ship traffic, and that was our opportunity to work fast!"

How to get ropes across the river was another challenge. The company leased a small tug to accomplish the task of pulling them. Instead of going underwater as anticipated, however, the ropes tended to float on top of the water, and the current carried them down the river. "Once we got one rope across the river, we made sure that we were never without a rope after that," Blackmon said. "We pulled the 'p-lines' lines, as we called them back in those days, three across, to make sure we kept one in the air all the time." Just to be safe – in case a rope broke – the crew simultaneously pulled a 5/8" steel-line, together with a 1 1/2" manila rope, and the conductor. This was quite a feat for the company's first work with LP&L, and the job had unusual circumstances right to the end.

"When our people got to the plant's side of the river," Blackmon explained, "the union employees at the aluminum company were really displeased that the work had been done by non-union workers." In protest, the Kaiser plant went on strike. Blackmon said that said Pike's crew members ended up wearing LP&L hats to work while they finished the job — a situation that Pike crews would encounter time and again in the future.

## CENTRAL DIVISION: OH, THOSE HEAT PUMPS

The company's expansion continued in the Carolinas with increasing numbers of crews and locations for service to Duke Power Company, which was itself experiencing tremendous growth in the '50s and'60s. Joe Maynor explained in *Duke Power Company: The First 75 Years* :

> In 1960, the company reported that it had more than doubled its generating capacity during the previous decade.... Except for a few minor setbacks, the nation's economy was in high gear as a new decade approached. Power demands were at an all time high, and the prediction for the future indicated that they would continue to increase rapidly in the 1960s.... Life, and industry, went on the Piedmont Carolinas. And the building boom which began in the early 1950s was getting louder and louder by the middle of 1964. During the first three months of the year, 18 new industrial plants ... had located in the Duke Power service area. At the same time, 77existing firms had expanded their facilities .... All of that expansion would require even more production of electric power. Duke Power was staying with the demand....[6]

The electrical industry, as a whole, was stimulating new areas of growth for the nation's utilities. "This was a time when the cost of electricity had been steadily going down," Joe Pike observed. He explained that many forecasters in the industry predicted that nuclear energy was just around the corner; and, with nuclear, electricity would become so cheap that it would not even be cost-effective to meter it.

*Duke Power Company: The First 75 Years* described the power company's participation in the activities that ensued:

> In 1958, the country was subject to the biggest promotional effort in the electric industry's history when a well-coordinated program

designed to get electricity into the home was announced. "Live Better Electrically" was launched with a multi-million dollar, nationwide advertising blitz. Industry promoters claimed that "there's something for everybody in the promotional plans — builders, retailers, distributors, contractors, bankers .... There was a national program to spur sales of new homes equipped with electric appliances.[7]

"Live Better Electrically" in the "All Electric Home" consequently prompted new ventures for Pike. Residential electric heating was virtually experimental when heat pumps came onto the market; and they were troublesome.

Floyd Pike worked closely with Duke Power executives during the heat pump's development. In a 1993 letter to Opal Pike, B.B. Parker, former president of Duke Power Company, explained:

W. B. McGuire (left) president of Duke Power Company, is shown talking with three officers of Floyd S. Pike Electrical Contractors Inc., (from left) Floyd S. Pike, president, Emmett W. Forrest Jr., treasurer, and A. Haywood Merritt Jr., vice president. Mr. McGuire, who was introduced by Mr. Pike, was the main speaker at the Mount Airy Chamber of Commerce banquet Monday night. (TIMES Photo)

*Floyd S. Pike, Emmett Forrest, Jr. and Haywood Marrett, Jr. meet W.B. McGuire, former president of Duke Power Company.*

Pike has had a broader relationship with Duke Power and Mill-Power Supply than just as a contractor doing electrical line-construction, and repair work. In the late 1950s and early 1960s, Duke was promoting the All-Electric Home. Pike Construction was an avid supporter of this activity, helping in promotional activities with other contractors, material suppliers, community leaders and related associations while working with Duke engineers in innovative and cost-effective ways to provide electrical service."[8]

As Floyd Pike continued to work with Duke Power in its endeavors, his efforts and his philosophy on which his company was based did not go unrecognized. He continued to form strong friendships with the power company executives with whom he worked. William B. McGuire, also a former president of Duke Power Company, wrote:

> Floyd believed in electricity and Duke Power Company. I believed in Floyd and was proud of him. He frequently came by my office when he was in Charlotte, and I looked forward to seeing him. He believed in good pay for a good day of work, and Pike Construction never backed away when the job was hard or the going tough. I well remember his struggles with the early heat pumps, those of other customers of Duke Power, as well as his own. It wasn't easy.[9]

During these exciting, but relentlessly challenging times, Floyd Pike reached a point of exhaustion that finally required a few days rest. McGuire' s appreciation for the efforts and long hours provided by the Pike company, and by Floyd Pike personally, extended into care and caution for the stress involved.

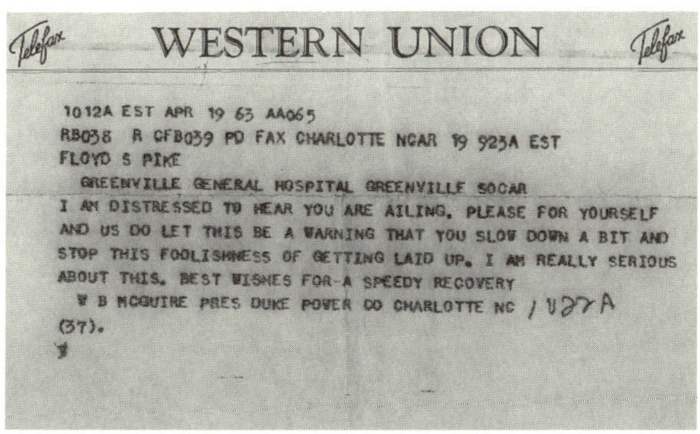

Telefax WESTERN UNION Telefax

1012A EST APR 19 63 AA065
RB038 R CFB039 PD FAX CHARLOTTE NCAR 19 923A EST
FLOYD S PIKE
GREENVILLE GENERAL HOSPITAL GREENVILLE SOCAR
I AM DISTRESSED TO HEAR YOU ARE AILING. PLEASE FOR YOURSELF AND US DO LET THIS BE A WARNING THAT YOU SLOW DOWN A BIT AND STOP THIS FOOLISHNESS OF GETTING LAID UP. I AM REALLY SERIOUS ABOUT THIS. BEST WISHES FOR A SPEEDY RECOVERY
W B MCGUIRE PRES DUKE POWER CO CHARLOTTE NC 1027A
(37).

*Telegram to Floyd Pike from his friend W.B. McGuire, former president of Duke Power Company.*

When Pike tapped into the heat pump market, Haywood Merritt was assigned their responsibility. He noted that the early heat pumps were large, one-piece units, and they were not very reliable. "I think that the first one the company installed was for Emmett Forrest," Merritt said.

Emmett Forrest confirmed that he, indeed, and astonishingly, did have the first "Weathertron" that Pike ever installed. At the time, he and his wife, Bobbie, had just begun building a house — their first house — built on a very limited budget. One night, the telephone rang, and it was Floyd was calling from a restaurant in Charlotte where he was in a dinner meeting with Duke Power executives developing the market for heat pumps. Floyd Pike's first words were, "Emmett, I want you to put a heat pump in your new house!"

Surprised by the call and, even more, at the request, Emmett Forrest said that when he began to ask about the cost differential, Floyd Pike repeated: "Emmett, I want you to put a heat pump in your new house!" Emmett Forrest said that he quickly understood that there would be no further conversation. "Yes, sir! We would love one!" He answered. "And, right away, we had the first heat pump in the entire county."

Emmett Forrest said that the "bugs" really had to be worked out on those first units. He recalled that the late Peter F. "Pete" Bondurant, who worked closely with Haywood Merritt in the Weathertron area, was often at his house, repairing and regulating this new apparatus. Not unlike the difficulties other service personnel have faced through the years, each time the heat pump has been modified with new technology.

At the time Pike was involved with Duke Power's development of the heat pump, Floyd Pike was singularly responsible for assisting the power company in a major cost savings. "Duke was trying to compete with the gas company on yard post lamps, but couldn't be competitive because they were too expensive to install," B.B. Parker explained. "Floyd came up with a cheaper way to wire and install these lights so that Duke could meet the gas competition for this service."

Opal Pike was also very involved in marketing during this time. She participated in many joint activities that Pike had with Duke Power promoting their "Electric Home". She often served as a guide at newly built model homes throughout the Carolinas. By the '60s, she had relinquished most of her office responsibilities and was Floyd Pike's full-time partner in all civic endeavors and their customer-related activities.

It is worthwhile to observe that although much of the technology developed during these years is now outdated, maybe even obsolete, it is evident in a personal note from B.B.

Parker that the friendships formed during these challenging years of Pike's joint efforts with Duke Power certainly have continued:

> We all like to win and hate to lose. Floyd was a real competitor. He was a great winner and a gracious loser. Some of our best times were when he and I played bridge with Opal and my wife — we, two rank-amateurs, played against our expert playing wives. Our occasional win was a cause for a great celebration.
>
> Floyd hated to lose at golf. He and I played about the same game. At Boca Raton in 1969, we were all even on the 18th green. Floyd missed his putt; and I made mine to win our customary $1 bet. You would have thought he had lost half his company! He found it so hard to give up that dollar, that I had him sign it. Enclosed is this dollar bill which says, "Floyd S. Pike 9/13/69 to B.B. Parker."

*Floyd Pike accepting the dollar prize from B.B. Parker.*

*Dollar presented to Floyd Pike by B.B. Parker, former president, Duke Power Company.*

> I am told that Emmett Forrest often appeared on the first tee complaining of a backache. Anytime this happened, everybody wanted him for a partner because they knew it was going to be a winning day.

In 1959, Pike Electric purchased stock in Hayden Electric Heating and Insulating Company, Inc., located in Pulaski. This move provided capital to Wilbur Hayden, a former Pike employee who was then establishing his own business, and moved Pike into the electric-heating market in Virginia.

## MORE EXPANSION AT HOME

While the heat-pump venture was struggling to climb out of infancy, another idea was born that eventually grew to become a new division for Pike: the Commercial and Industrial Wiring Division. Glenn Pike was a foreman in 1956 when back trouble took him off the job. He said that he remembered Floyd Pike saying he would like to get into the electrical wiring business someday. So while unable to work, Glenn Pike studied for the unlimited contractor's exam and qualified for a commercial wiring license. "When I returned to work, I talked to Floyd and told him what I had in mind. Floyd's reply was, 'If you want to, we'll start this business tomorrow morning!'" While not exactly the next morning, the division did officially begin operation in July '60.

"It took a little while before we zeroed in on what we needed to do," Glenn Pike said. He said that he didn't know anything about estimating at first and that learning to figure the cost of the jobs, materials, and labor was a "pretty good little chore!" Although it was slow in the beginning, the division began picking up a job or two. "The first thing we knew, we were getting a little more and a little more, and we began doing pretty well."

Jim Watts joined the Commercial Wiring Division in '62 and has been a part of its success for over thirty-four years. While Glenn Pike did the bids, Jim Watts handled the purchasing, billing, and the other office tasks. Later on, Watts supervised the division's residential work. He even sold base-board heat by phone, and door-to-door, from his own vehicle. "We did whatever was needed," Jim Watts said. "And as a result, we saw the Commercial Division continue to grow."

As the work increased, William Wagoner, Ray Gwynn, Roy Hiatt, Gray Noonkester, Curtis Dickerson, and others, were out on the jobs. Pauline Pyles came to work in the office, along with Paula East.

To join together the efforts in developing the heat pump with those being made in the Commercial Wiring division was a fitting course. Glenn Pike explained that one thing led to another:

> After a while, I suggested to Floyd that we get into the air conditioning business, get us a sheet metal shop, and start doing the fabricating, the duct, and the insulating — the full line. He decided that might be a good idea. So we bought Charlie McCraw' s sheet metal shop and hired Charlie. The first job we bid was the Mayberry Mall in Mount Airy. We were successful in getting that job with the late John S. Clark and his company. He was a very, very nice man to work for, and we did a lot of jobs together.

The Commercial and Industrial Wiring Division provided service to Hanes Corporation in Winston-Salem, where the Pike crews reported to the Superintendent of Maintenance at the time, Joe D. Patterson. In the course of working with Pike's people there, he was approached by Floyd Pike who suggested that Patterson talk to the Pike company if he ever decided to leave Hanes.

In 1967, Joe Patterson moved his family to Mount Airy, joined the Commercial and Industrial Wiring division, and assumed the responsibilities of industrial air conditioning that

Glenn Pike was in the process of expanding. Patterson was indeed well-qualified, having been in charge of installing and maintaining 9,000 tons of centrifugal air conditioning equipment while at Hanes.

Shortly after Patterson's arrival, the commercial operation was moved into larger quarters in the Allandale Building on South Street in Mount Airy, where the division purchased and installed a sheet-metal layout table for fabricating duct work and a pocket lock machine. "An office was constructed," Joe Patterson said, "and then we started bidding on commercial installations!"

The commercial work began small with only the crew that had worked for Charles McCraw in metal fabrication and three mechanics for equipment installation who were already employed with Pike Electric. It was not long before they expanded to two service crews and hired Ed Fore.

Fore, now retired — unless the company needs him — is the father of Don, Jerry, and Gary Fore and the grandfather of Kevin, Jamie, and Billy Fore, three generations of Pike personnel in the Commercial Wiring Division. Ed Fore is also the father-in-law of HVAC dispatcher Tim Flippin

The jobs increased under Joe Patterson's direction, and another long-time employee, Butch Riley began working for this division. "Although I had a lot of engineering training in my portfolio, I was very 'green' in the area of HVAC and contracting," Butch Riley said. "I don't think I was of much benefit to Joe at first. He was very patient with me and taught me a tremendous amount about building load calculations, on-site surveys, equipment sizing and estimating." Riley said that all the while, "Joe Patterson stressed excellence in customer service."

The division began getting more challenging assignments. One, in particular, was a Holiday Inn in Lexington, North Carolina, where Donald Fore was in charge of equipment installation. "Donald Fore is the best air conditioning mechanic I have ever known," Joe Patterson said. "Before this job, Fore had been in charge of residential unitary and split-system pump installations, and he was really nervous about installing larger air conditioning equipment." Patterson reminded Donald

Fore that whatever the size or brand of a system, the same type of control equipment protected it. As the Holiday Inn job progressed, "Donald called to tell me to bring on the equipment," Patterson said. "He was no longer intimidated by *anyone's* equipment!"

The Commercial Division's work force grew from seven men to twenty-five in just over a three-year period. During this time, the department was doing practically the total electrical and air conditioning installation for local and state-wide building contractor and long-time friend of everyone at Pike, Jerome Samet, whose work involved private residential building, apartment complexes and light commercial construction at the time.

Joe Patterson recalled a variety of memorable jobs that included installation of residential heat pumps in condominiums at Doe Run, the Ground Hog Mountain resort near Meadows of Dan, Virginia, as well as installations in some magnificent private residences. He added that other early unforgettable jobs ranged from service for centrifugal refrigeration compressors to installations of some exceptionally large heating and air conditioning systems that included the Spencer Manufacturing Company's system in Mount Airy and a system for the Champion Dishwasher Plant in Winston-Salem. Jobs were as varied as: the clubhouse at the Tanglewood Country Club near Winston-Salem; movie theaters; banks (at the time when branch bank building was just beginning); a newspaper office; supermarket chains. It was during this time that Pike installed the first two, of many, heating, air conditioning, and ventilation systems for the Hardees Restaurants.

"There are many other jobs, both large and small, that we installed in the eight years that I was with Floyd S. Pike Electrical Contractor, Inc.," said Joe Patterson. "One could stand in the middle of Mount Airy and count about 800 pieces of equipment that the company installed. I am very proud of this fact."

In 1974, the Pattersons left Mount Airy and the Pike company to open their own business in Wilmington, North Carolina. In recalling his years with Pike, Joe Patterson summarized: "This period of my life is one of the highlights.

Knowing all of the Pikes, Lawrence Allen, Zack Blackmon, Emmett Forrest, Haywood Merritt, the guys in the garage, and all the others, is ever in my thoughts."

*Jim Watts.*

*Don Fore.*

*Gerald Francis and Doug Perkins.*

*Butch Riley and Pike Electric HVAC Project Manager, Gary Crews.*

*Doug Pauly, Gerald Francis and Gary Culler.*

*It all started the second time for me in November 1965.*
*I was home from Raleigh on a two-week vacation,*
*when I saw Mr. Pike at church.*
*He asked me to take a trip with Zack Blackmon,*
*Sr., to Baltimore and Washington the next week.*
*Just before we left that Monday morning,*
*Mr. Pike handed me a piece of paper with instructions*
*not to open the note until Zack and I*
*got well up the road on our way to Baltimore.*
*He said he had written on that piece of paper*
*how much money the company was going to pay*
*me to come to work.*
*I told him that I wasn't looking for a job;*
*but he insisted that I take the paper anyway*
*and make up my mind later.*
*The outcome of what he had written,*
*of course, is obvious. Here we are now,*
*thirty years later, on a trip that was only going to take four days!*
*—Claude Brinkley*

15

# Making the Work Go Smoother

## Safety

With more employees working in more areas on jobs that often presented amazing challenges, such as the Mississippi River crossing, the potential for accidents increased. Safety has always been of major importance to Pike, but in the '60s the need for additional concern was recognized, and an infant safety department was set up. "Workers' Compensation costs may have played a roll in initiating firmer safety polices," according to reports from Zack Blackmon, Jr., presently in charge of the company's human resources division, including safety, and A.L. McAlexander, former director of safety. "Insurance costs were directly proportional to Pike's accident cost: fewer injuries. lower costs."

Who could have known better about the potential for accidents than Joe Fulk who had just climbed to "new heights" in Louisiana. He was named the company's first safety director in 1963.

Called the "Safety Man," Joe Fulk incorporated hs duties as Safety Director into his other responsibilities at the time, which ranged from field mechanical work to transmission construction supervisor. He had the qualifications for this new job through his volunteer service to the local rescue squad. He was well-trained in first-aid and in administering artificial respiration. "He really got the department off to a good start." Blackmon and McAlexander agreed.

*Floyd Pike and Jack Farris as Farris accepts life - saving award for heroism in Winston-Salem from Pike's first Safety Director, Joe Fulk.*

When Joe Fulk's other responsibilities increased, W. E. "Eddie" Bondurant took over the position of safety director. Eddie Bondurant is the son of long-time Pike employee, Pete Bondurant, now deceased. During his tenure as safety manager, Bondurant assisted management in putting together the company's first written employee safety manual that contained many of the rules which are still in force today.

## EXPANSION ENLARGES OFFICE PROCEDURES

This period of expansion affected all aspects of Pike's operation. The men on the crews were not the only Pike employees who were proving that they were both adaptable and capable of meeting the specific requirements of each individual customer. Likewise, the office staff efficiently accommodated each customer's billing system, whether unit-priced or hourly-contract. And did so: pre-PC's!

"The office staff and garage always regarded themselves as support units for the men in the field," Emmett Forrest said. "When the numbers and requirements of the men in the field accelerated, for example, during storm trouble, the small office staff — like the garage — met the challenge."

*The first storm I ever worked was Hurricane Hazel.*
*When that storm rolled into North Carolina in the fall of '54,*
*Mr. Pike gathered us up and sent crews to work there for*
*about three days. Then in '58, we gathered up everybody we*
*could find and went to Baltimore, and we worked the ice*
*storm there for about three weeks.*
*Hurricane Betsy hit the coast of Louisiana in'65.*
*We gathered up everything that could be released*
*and traveled south.*
*We worked that storm for about three months.*
*— Richard Simmons*

16

# Storms of the '60s

Storm trouble demands tremendous efforts from all departments of the company. In the early 1960s, ice caused tremendous damage in Georgia, and Pike's crews provided storm assistance there in '61 and again in '63. However, the company had never before confronted disaster of a greater magnitude or dealt with a storm location quite so far away, until Hurricane Betsy came out of the Gulf of Mexico in 1965. The call for help from the Louisiana Power & Light Company — that resulted in an industry-wide awareness of Pike's reputation as a No.1 storm response team — presented a real logistical challenge.

When the storm hit, Joe Fulk was most familiar with the area around New Orleans, and he was immediately dispatched to Louisiana to begin organization. Claude Taylor, Pike Electric assistant vice-president, headed south to Louisiana from Virginia. Wade Martin, Ralph Anderson, and Blaine Bard quickly moved their crews from North Carolina. Before the storm work ended, Pike had almost two hundred men involved in the restoration.

Glenn Dobson, who had just been promoted to supervisor, went directly to New Orleans from Georgia. "This was my first experience with a hurricane; and oh, what an experience!"

When he arrived in New Orleans, Glenn Dobson took his crews to the operating center of LP&L's East Division to await the arrival other Pike crews. Dobson said that when he received the first call there from Floyd Pike, he quickly learned what his new job was. "Floyd said for me to handle everything that was needed by all of our Pike crews that were working out of LP&L' s east-side center during the mornings. Then he told me to get across the river to Gretna and take care of all of our crews operating from LP&L' s west-side office in the afternoon."

Dobson said that along with everything else involved with the restoration, he was dealing with "no power, no water, no rooms!" The power company had booked quarters for all of the crews that were there to assist. "But we worked so late that when we got to the motel, our rooms had been given to others," Dobson lamented. "It took me until 3:00 a.m. to get the crews a place to sleep, and when we finally did get some rooms, we had to carry water from the motel's swimming pool!"

The entire area was devastated from New Orleans, south. The utilities in the area had left only the Little Gypsy Steam Plant in operation, in order to have one plant from which to start the other generating units as they were needed after the downed lines were rebuilt. Pike crews were responsible for rebuilding both transmission and distribution lines. However, like all storms, the restoration following "Betsy" had some unusual twists.

Logistically, Pike's personnel encountered some real surprises. Joe Pike was handling the payroll; just getting paychecks to the crew members in some of their locations was a feat. "Often, the only way I could find them, and then get to them, was by helicopter," he said. "They were working all over the place!" Joe Pike delivered checks, picked up time sheets, went back into New Orleans to get Pike's pay from LL&P, and then flew back to Mount Airy. When the invoices and the checks were processed and ready to go, he would head back to New Orleans. "I shuttled back and forth about every seventy-two hours from Mount Airy to New Orleans — that was typical for everybody," Joe Pike said. "All of us were shuttling back and forth doing one job or the other for about three months."

All the while that Joe Pike was trying to get the payroll to the crews, Pike's transportation personnel were hauling equipment and supplies to them. Joe Pike said that Ralph Puckett must have been on the same shuttle schedule as he. "It seemed like we met on the road every trip!" And many trips were necessary. "Besides getting what was needed there to start with," Pike continued, "the conditions were pretty bad and equipment required a lot of repairs and replacements."

The logistics for the drivers were grueling and demanded not only skill, but patience as well. Due to the distance, Pike trucks ran as heavily loaded as possible. However, when they reached the Mississippi state-line, they encountered stringent weight-restrictions. Half the truck's load would have to be removed and left waiting, before the truck was allowed to drive across the state. When the driver and his rig crossed into Louisiana, the truck would be completely unloaded. Making the return run across Mississippi empty, the driver would pick up the waiting load, again drive across the state to retrieve the remainder of his load, and finally proceed to his destination.

The restoration itself had its own challenges. One particularly hard-hit area where Pike crews were needed was around Buras, Louisiana, thirty miles down the Mississippi from New Orleans, where Louisiana Power personnel took Zack Blackmon, Sr., by helicopter to survey damage and organize a work plan. "The crews were to repair a diesel-powered plant that had gone completely underwater," Zack Blackmon, Sr., explained. "It had nine big diesel generators in it — all totally gone. The power company's employees had already been working on it for days and had one engine almost ready to start up, but they had no nowhere to send the power. All lines were down."

On the river, LP&L had docked a ship with a power supply. "We were to take the ship's 4160-power, via an underground cable, to a substation about 300 feet away, where it was transformed to 12,000," Blackmon explained. "Then we were to start building power lines, literally, from scratch."

Zack Blackmon, Sr., said that Pike trucks were about the only thing on the road going to Buras that day, because the area was hit so hard that the residents who had returned to their homes, had found nothing, and just left. "They were all leaving, heading north."

When Zack Blackmon, Sr., Ralph Anderson, and Blaine Bard arrived in Buras with their crews, they found living conditions that were really rough. They were housed in a school with no power. All the cooking was being done on GI oil stoves. "The whole place was right scary looking!" Zack Blackmon, Sr., remembered. "Within a couple of days, we had power into the schoolhouse which made things a lot better," he added, "and we soon had some pumps running to supply water – undrinkable, but it did allow us to get cleaned up a bit."

The excitement didn't end then, however. "We hadn't been there even ten hours," Blackmon remembered, "when we received word that another hurricane was reportedly heading our way!" He recalled that Ralph Anderson looked at him and said, "I don't believe I can keep our people down here. With that kind of report, this place is liable to totally wash away!"

Ralph Anderson's concern was a legitimate one, because in that area, the Mississippi is diked 30 feet high. "Even the ship that we were to work from was 30 feet above us there!" Zack Blackmon, Sr., said. "We stood on the ground and looked up at it!"

Fortunately, the hurricane turned away, and the area received only a "tough rain." "It didn't help morale any," Blackmon commented, "but those boys stuck it out down there. Blaine Bard stayed there for several weeks with his crew, building a line up and down the river."

Before the restoration was completed, Gene Lancaster

called Pike and asked for additional help with a difficult situation that had occured. In the swamp and marsh and devastation, helicopters were being used routinely for reconstruction. In attempting to lift a pole, a pilot working for LP&L, accidentally lifted a static conductor. When the pilot realized his situation and the danger of stalling the plane, he tripped the line, and it fell to the ground near a Louisiana Power and Light crew. Understandably frightened, the LP&L crews walked off the job and refused to return to work.

After giving the helicopter force strict instructions and obtaining assurance that they would operate according to Pike's safety standards, Claude Taylor and his crews used the helicopters with no problems and set over 300 poles. They built mile after mile of transmission lines, enabling LP& L to get one station, after another, on line.

Louisiana became an area of real expansion for the company. Following the clean-up, a contract for unit-distribution work was obtained. Joe Fulk, who later became a Pike assistant vice-president, remained there with crews through several years of work.

*Roy Thomas on the job in Louisiana.*

## STORMS OF A DIFFERENT NATURE

Turbulence of quite a different nature caused further expansion for the company. Word of Pike's willingness to tackle the most challenging tasks in the electrical industry reached Kentucky Power Company, which was embroiled in a strike and needing assistance. "Waldo LaFon, Kentucky Power's chief executive officer called and asked if Pike could work for them," Haywood Merritt said. "We did! Although, we couldn't be very productive."

The strikers were harassing the Pike crews in every way possible. At one point, they even called the sheriff and had Pike's trucks impounded for trespassing, claiming that the company's crews were loading Pike's equipment on their property. After delays that were caused by legal procedures and having to rely on the aid of the local sheriffs for protection at work, Waldo La Fon had had enough: "Put those men on *our* trucks, and get them back to work!" he ordered. And so, Pike crews went to work on Kentucky Power Company's trucks.

The harassment didn't end there, however. After the trucks were loaded and en route to the job site, the strikers slowed traffic to a crawl. Added to this inconvenience, the union workers sprinkled the truck parking lots with tacks, and they walked picket lines around individual poles where Pike's linemen were working. This created tension and physical stress on Pike's personnel. "We decided that it would be only fair not to keep the same people up there for long stretches of time," Haywood Merritt said. "We worked in rotation. One group of linemen worked for a week only, and then they returned to their regular assignments when new crews arrived to carry on."

The extraordinary effort made by Pike's personnel at that time provided immediate benefits to the customer, and long-lasting benefits to Pike, in return. Eventually the strike was settled," said Haywood Merritt, "and Kentucky Power Company negotiated a unit-price contract with us." He added, "We have been working on their property continuously since then."

"During this time, we got to be on a very close and friendly relationship with Kentucky Power's top management." Merritt said. "I remember being in Waldo LaFon' s office one time when he was on the phone telling a chief executive of the utility's parent company, American Electric Power: 'Do not award this unit-price contract to anybody but Pike unless you talk to me first!' "

Pike has built a reputation for quality service and a "can-do" method while working on many jobs that were carried out in the toughest of situations. For example, it was Pike's tenacity, while working for Virginia Electric Power Company at a time that VEPCO, too, was experiencing a strike, that led to many years of work. "Following the strike," Merritt recalled, "Pike probably had eighty men in and around Richmond and Norfolk for eight or ten years." He said that the "added benefits," with VEPCO and Kentucky Power, as with so many other customers, were the close personal friendships that began then and continue today. "Those are the kind of relationships that we developed," he said.

## RESCUING A COMPETITOR

There is a corollary to the friendships and the working relationships that resulted from Pike's reputation for help in times of hardship. Included in this period of Pike's history is an extraordinary episode. However, the parallel that occurred was not with a customer, but with a competitor.

Floyd S. Pike, Floyd S. Pike Electrical Contractor, Inc., and the C.W. Wright Construction Co., Inc. of Richmond, Virginia entered into an unusual contract August 1, 1961. Jack Wilson, a CPA with Strand, Skees, Jones & Company, provided the accounting service that was required for Pike's endeavor:

> It is my understanding that Floyd S. Pike was personally acquainted with Charles W. Wright, who was president and owner of C.W. Wright

Construction Co. Mr. Wright informed Mr. Pike that he was in financial distress, and that he needed help or else he was going to be forced into bankruptcy and out of business.

Floyd Pike, for whatever reason, undertook to guarantee, personally, himself, and to pledge the credit of Floyd S. Pike Electrical Contractor, Inc. to assist C.W. Wright Construction Co. in its efforts to restore itself to solvency. Floyd required that he have a very significant and substantial say-so in the running of C.W. Wright Construction Co., as long as he was serving as a guarantor. Floyd undertook to rebuild, reconfigure, and to restore morale to this company. He literally used his own influence to help his competitor secure contracts. It was through Floyd's influence that C.W. Wright was able to become an accepted contractor with the various power companies. All the while, Floyd was still working with some of the same power companies himself.

Not only was Floyd Pike very instrumental in rebuilding the Wright Co.'s equipment and truck fleet, but he was also very influential in rebuilding the management structure of the C.W. Wright Construction Co. as well.

The arrangement was in effect for roughly five years, and during that time, the company was brought back to a profitable status, and had regained its creditability among its creditors. C.W. Wright Construction Co. was successful in restoring its credit and relieving Floyd Pike and Floyd S. Pike Electrical Contractor, Inc. from all obligations in the summer of 1966. The companies have maintained a professional relationship and have functioned as competitors since 1966.

> The relationship between C.W. Wright Construction Co. and Floyd Pike, personally, and Floyd S. Pike Electrical Contractor, Inc., was an unusual one in which a competitor, who happened also to be a personal friend, pledged his credit and the credit of his company to assist his competitor back to financial health.
>
> The mission was accomplished. But I doubt seriously, and this is a personal opinion, that you will find many other situations such as this in the business world. I have been in public accounting for over forty years, and this is first time that I have ever seen a competitor help another back to health.

Simply, and succinctly, Opal Pike explained the circumstances: "Charlie was in trouble, and he was our friend. You remember, right from the start, many people helped us along the way."

*I came into the company full time in '65.*
*For the first few weeks, I followed Warren Hill, Zack Blackmon, Sr., and Horse Maynor around to see what kind of work the company was doing — downtown and residential underground, and, of course, our regular distribution work.*
*I didn't get involved in any transmission work at that time.*
*One Saturday morning in late December,*
*I came into the office, and Mr. Pike met me at the door.*
*He told me that my equipment was across the street,*
*and that I would be going to Alexandria, Virginia,*
*on Monday morning to work in Haywood Merritt's division.*
*I'll never forget that day, because it had been ten years*
*since I'd worked on a line crew!*
*I didn't know what an 052 tool was, or a press.*
*I'd never seen a bucket truck work, or an hydraulic-line truck at work.*
*I thought to myself: "Man! What is he doing to me!"*
*But the next Monday morning, I found myself working in Alexandria as a foreman of a seven-man line crew.*
*What a shock!*
*I quickly figured out that if Mr. Pike and Haywood thought I could do it, then I better do it. I've never regretted that year's training.*
*During that time, I learned what it was like to work in all seasons of the year.*
*I learned what people can do in a lot of situations*
*that you sometimes only hear or read about.*
*Since then, I've always been able to say, with some authority,*
*that I understand what a lineman, a foreman, or an underground person is thinking about*
*— what might be going through their minds —*
*when they make the decisions to do the things they do.*
*I hope that all of our young people appreciate the training we provide.*
*They are the future managers of the company.*

*—Claude Brinkley*

# 17

# Growing Pains in the '60s

While the Pike Company continued to build regular distribution systems in residential areas for many customers, the company was venturing into new endeavors: major transmission construction and underground work. "Our transmission work was a very exciting area with lots of high-visibility jobs that required an enormous amount of equipment — very big, very specialized, very expensive equipment," said Joe Pike. "At our peak in transmission, we were equipped to handle almost any job, including EHV work of 500 & 765 KV lines. We built major parts of the Southeast power grid during the late '60s and early '70s."

At the same time, underground was just becoming a very big movement in the electrical industry. Underground work involved converting overhead lines to underground for the entire downtown areas in some cities. This work also included a lot of major underground duct and duct-bank installation. In the course of this work, Pike employees became amateur archaeologists in some areas — their digging turned up all kinds of "finds" — even centuries-old streets with abandoned street car tracks! Underground work created a major expansion in the company's growth. Growth, for the Pike company, meant "challenge" — a word that Floyd S. Pike defined as "opportunity"!

"Our 'growing pains' were good ones," Joe Pike said. "They were persistent, and they were increasing. We needed more men with more equipment — and more capital for both. And often in a hurry! Financial initiatives were imperative."

## FINANCING WITH FRIENDS

"Well, later on, Floyd needed a connection with a bank that had a larger lending limit than the banks at Mount Airy had at that time," said F. D. Forrester. Since the company had continued to buy trucks from Forrester at Wilkes Dodge, he was well aware of Pike's rapid growth and the need for additional capital for that rate of growth to continue. He introduced Opal and Floyd Pike to Edwin Duncan, Sr., president of the Northwestern Bank at North Wilkesboro. "Mr. Duncan was very well impressed with Mr. Pike," said Forrester. "They immediately hit it off as good friends."

The relationship that formed between Duncan and Floyd

Pike began with a handshake; it matured within a context of mutual admiration. The two men had ultimate faith in each other. Opal Pike related that Edwin Duncan, Sr.'s advice was adhered to by her husband, "as if given from his father." For the next twenty years, the significance of this friendship affected the course of the company's history of growth and brought about memorable personal experiences for Floyd and Opal Pike as well.

*Floyd Pike and Edwin Duncan, Sr.*

How important was the Northwestern Bank to the Pike Company at this time? A statement from a '65 edition of the *Winston-Salem Journal and Sentinel* explains:

> A good deal of the growth that has occurred in the Northwest has come about because Ed Duncan was on hand to prod it along. …
>
> There are 44 industries in Northwestern North Carolina which are there because Ed Duncan and his country bank had the foresight — and the cash — to help them get over the early humps."[13]

*The remarkable thing is that it is the crowded life*
*that is most easily remembered.*
*A life full of turns, achievements*
*disappointments, surprises, and crises*
*is a life full of landmarks.*

—*Eric Hoffer, philosopher*
*Reflections on the Human Condition (1973)*[14]

18

# Floyd S. Pike in the '60s

## Dedicated Citizen

Edwin Duncan, Sr., was responsible for providing the Pike company with financial backing in the '60s and '70's that was commensurate with the company's extraordinary growth. At the same time, he also encouraged, and sometimes was a part of, Floyd Pike's numerous civic, industrial, and political endeavors. For example, at Edwin Duncan, Sr.,'s, request in the mid-'60s, Floyd Pike began serving on the executive committee of the Northwestern Bank's executive board of directors. He also served on Northwestern Bank's local board of directors in Mount Airy.

During this time, Floyd Pike was working with other leaders in his hometown to strengthen Mount Airy's Chamber of Commerce. His civic and industrial efforts were recognized in 1966, when he was presented the Chamber's "Citizen of the Year" award. He worked with the National Chamber of Commerce as well.

Floyd Pike enjoyed golf, and, in the '60s, he served as president of the Mount Airy Country Club. He was also chairman of the board of directors of the Surry Community College Foundation. He served as president of the Mount Airy Youth Foundation. He was president of a small textile mill, Mount Airy Hosiery. And, he served on the board of directors of the New Orleans Baptist Theological Seminary.

During the '60s, Floyd Pike spent a great deal of time in Raleigh, North Carolina, where a major legislative effort was underway at that time for the "Utility Territorial Bill" to become law. Kelly Hutchens, then manager of Surry-Yadkin Rural Electrical Membership Association (REA), cited Floyd Pike's successful efforts in bringing together "many interested parties" who were necessary for the "passage of a law that as been very beneficial to the concerned utilities, their customers, and members."

*Floyd and Opal Pike in the '60s.*

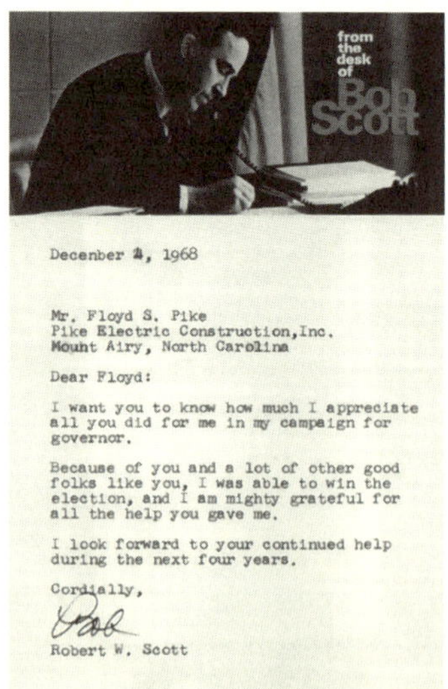



Decenber 4, 1968

Mr. Floyd S. Pike
Pike Electric Construction,Inc.
Mount Airy, North Carolina

Dear Floyd:

I want you to know how much I appreciate all you did for me in my campaign for governor.

Because of you and a lot of other good folks like you, I was able to win the election, and I am mighty grateful for all the help you gave me.

I look forward to your continued help during the next four years.

Cordially,

Bob

Robert W. Scott

*Floyd Pike and Govenor Dan. Moore of North Carolina.*

*Pat Huff, (left) former vice-president of Duke Power Company with Floyd Pike in the '60s.*

*Baseball great Yogi Berra with Floyd Pike.*

*Floyd Pike was president of Mount Airy Country Club in the '60s.*

*Left to right: Emmett Forrest, Nick Shoaf, Zack Blackmon,Sr., R.E. Huggins, Jim Sloan, Floyd Pike, Haywood Merritt, Jr., unidentified guest, Roy Thomas, Walt Reeder, unidentified guest, local Mount Airy golfer, Wendy Williamson, Bob Abercrombie, and Lawrence Allen.*

# MOUNT-AIRY SURRY COUNTY AIRPORT

Floyd Pike loved flying, and he well understood the importance of flying in the logistics of his own business and to other businesses as well. He realized, too, that an airport was essential for any community's industrial and commercial future. Mount Airy's only airport facility at the time was a grass strip. Floyd Pike served as chairman of a group of Surry County community leaders that included: John Banner, Jr., Howard Woltz, Jr., Raymond A. Smith, and Dr. Louis Spillman. Under Pike's leadership, the challenge of bringing airport facilities to Mount Airy was accomplished. With grants from the Appalachian Regional Commission and the FAA, along with federal funds — and, without any local taxpayer money — the runway was built. Tireless efforts ensued to solicit private funds from individuals and local firms for the project's completion. The facility finally became a reality in 1969, after almost ten years of struggle. "The Mount Airy-Surry County Airport would never have been built without Floyd Pike's untiring efforts," said John Banner, Jr., a Wilmington, North Carolina businessman, who is originally from Mount Airy, and who served as vice-chairman of the airport committee. On its completion, the Mount Airy-Surry County Airport was given to the town and the county jointly.

The airport's existence was soon threatened by poor management, and the operation faced bankruptcy little more that a year after it began. Meanwhile, the Pike company's plane, which was based in Mount Airy, had proved to be an expedient tool not only for management, but also for transferring line and fleet personnel — a flight home was especially welcome to a crew after working long hours on storm trouble. And the plane often transported specialized equipment to a job site where it was needed immediately, if not sooner. With the growth that the company was experiencing, job locations were more distant and scattered. The plane allowed more people to be in more places with less time-loss in travel.

Many employees, customers, and suppliers will fondly recall Pike's chief pilot for many years, Maynard Snow. He was usually accompanied by Jerome Samet, who never refused to leave his own business any time that he was needed to fly as Pike's co-pilot. These two men were not only the corporate pilots, they were representative of the Pike company's professionalism. When weather, especially storms, presented particular challenges to them, their usual disclaimer was "that they were only called, when the weather was too bad to drive!"

Distance and mobility were the deciding factors that caused Pike's management to purchase the local airport operation and to operate it as a wholly owned subsidiary. Maynard Snow assumed the additional duties of airport manager until the company discontinued the operation in 1975.

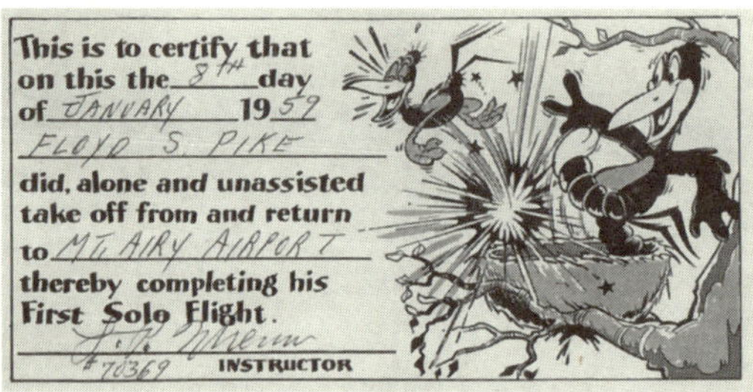


*Floyd Pike loved to fly.*

*Floyd and Opal Pile with one of their first planes, late '60s.*

*Floyd Pike and John Banner, Jr. survey the future Mount-Airy Surry-County Airport.*

DIGNITARIES FORMING THE OFFICIAL dedicatory committee included (l-r) David McGavock, John Banner, the Rev. Tom Bagnal, Hugh P. Sherrill, E. R. Short, Raymond A. Smith, Floyd Pike (at microphone) Howard Woltz, Maynard Beamer, Boyd Tilley, John Simmons, Bowman George, Grady Harmon. —NEWS Photo

# Local Airport Authority Gives Facility To Town And County

U. S. postmen are supposed refuse to allow the weather under them, no less staunch in determination were officials the Mount Airy-Surry County Airport Authority Sunday.

With the gala trappings of a holiday about, the airport commission endured searing heat Sunday afternoon to stage, on sharp schedule, formal dedication ceremonies of the new local airport. A long trailer parked inside a service hanger served as a platform for the occasion.

As the aluminum sides of the hangar fairly crackled in the heat, the several dozen citizens, pilots and well-wishers gulped cold drinks from a nearby refreshment stand and listened as Floyd Pike, chairman of the Airport Authority, presided over the brief rites.

After the introduction of several dignitaries on the platform, Pike turned to Howard Woltz, secretary of the authority, for a resume of the airport's turbulent beginnings.

After a review of the trials and tribulations the group experienced in getting the air facility established, Woltz reported that today the airport is "a safe and well financed, conservatively operated" enterprise.

He formally presented the airport to the Town of Mount Airy and Surry County, with mayor pro-tem Bowman George accepting, and John Simmons, chairman of the board of commissioners, doing likewise for the county.

Because the facility is owned by the local municipalities, Woltz noted that contributions to it are tax deductible, a factor he emphasized for his audience.

The Mount Airy Jaycees sold boxed lunches during the noon hour. There were numerous planes which arrived at the facility to participate in the dedication. A "fly-in" of antique aircraft failed to draw as heavily as the authority members had hoped. They awarded a case of oil as a prize to the oldest plane present, a 1940 model Waco biplane boasting two open cockpits. It was the subject of close scrutiny throughout the day as the hundreds of visitors to the airport wended their way around the airport, eying both new and old planes.

David McGavock, operator of the airport, was kept busy during much of the day operating an air taxi service, taking numerous spectators for aerial views of the airport and countryside. A helicopter at the field did similar duty for several passengers.

Pike indicated that his group will continue to press for improvements and additions to the facilities at the airport. Woltz said the airport is operated at a small profit and that future profits should be enough to finance some of the additions the Airport Authority envisions.

*Floyd Pike, chairman of Mount Airy-Surry County Airport at dedication ceremony.*

## NORTHWEST NORTH CAROLINA AREA DEVELOPMENT ASSOCIATION

Floyd Pike was a leading member of the Northwest North Carolina Area Development Association during its most productive period, the 1960s. The organization's purpose was to facilitate progress in industry, education, agriculture, travel and tourism; which, in turn, generated community improvements and prosperity for the state's entire mountain region. North Carolina's "High Country," Pilot Mountain State Park, and Stone Mountain Park are three of the best known and most widely enjoyed results of the efforts made by this organization during these years. Many new, diversified industries, both large and small, settled in northwestern North Carolina during the time. Floyd Pike was elected president of this organization in 1968.

With his confidence, his common sense, and his perseverance, Floyd Pike's participation influenced any group's endeavors. He projected a vitality and a consummate belief in one's ability to achieve any worthy goal, not just into his own business, but into anything he attempted. His personal characteristics and his philosophy are evident in his final address as president of the Northwest North Carolina Area Development Association:

> As we pause at another year's end, we look back briefly to see where we have been, to measure the progress we have made as an area development association. However, the Northwest North Carolina Area Development association is not a "look-back" organization. In fact, we are just beginning to grow....
>
> This is a mountaintop moment, a time to look out and see anew our opportunities and to rededicate ourselves to the tasks before us. There never has been a need for the word, contentment ....
>
> Community effort cannot be replaced by any substitutions. Improvement or betterment for our people is our own job.

STATE OF NORTH CAROLINA
EXECUTIVE DEPARTMENT
RALEIGH

DAN K. MOORE
GOVERNOR

February 13, 1968

Dear Floyd:

You and the other officers and members of the Northwest Development Association made our week end visit to your section a most pleasant one in every respect. We do appreciate all of your many kindnesses and the opportunity to have seen the progress which is being made in your part of the State, much of which, of course, can be attributed to you and your Association.

We particularly enjoyed the reception and dinner at North Wilkesboro and the reception at Hound Ears. The next time you are in Raleigh, please stop by so that I may extend to you my personal thanks.

With best wishes, I am

Sincerely,

Dan

Dan Moore

Mr. Floyd S. Pike, President
Northwest North Carolina
Development Association, Inc.
Post Office Box 868
Mount Airy, North Carolina

DM:adl

*Floyd Pike served as president of the Northwest North Carolina Area Development Association in 1968.*

*The next Rural Electrical Membership Cooperative*
*that I called on was the Sawnee EMC at Cumming, Georgia.*
*The manager was Frank Chancey.*
*"I'd like for you to get me a crew together," he said.*
*When I left that co-op office, Oh, Boy! I was a happy man!*
*— Roy Thomas*

# 19

# A Southern Division

Indicative of a decade highlighted by personal and corporate achievement, Pike's large wall map was marked with a central pin that designated the source: the company's headquarters in Mount Airy, North Carolina. Vectoring in all directions, color-coded pins, representing crews and their locations, were constantly systematized by Pete Bondurant.

Toward the mid -'60s, Roy Thomas "went South" in earnest and began developing Pike's Southern Division. He moved his family to Monroe, Georgia, where he established headquarters and began calling on potential customers, mile by mile. "Even after Mary and I re-located there," Thomas said, "I was driving about 75,000 miles a year on the average for the next ten years in Georgia."

When work slowed down, Thomas would "knock on more doors," and when work picked up, he would then concentrate on getting more crews started. "We also strengthened our reputation there when we provided help during the big ice storms in Georgia during the early '60s," Thomas said. "And whenever there was a co-op managers' meeting, we would maintain a hospitality room which gave us a lot of recognition among the people around there."

Pike's hospitality rooms have been the source of many friendships. "Often in our hospitality room, "Roy Thomas explained, "we have had an opportunity to work out something that might be troublesome before it ever becomes a problem." Staffing a hospitality room affords Pike's personnel additional opportunities to meet with customers and suppliers, discuss industry-wide occurrences and equipment findings. Potential opportunities for work are often identified in conversation through this forum. For many years, on many occasions, Pike has worked jointly with Altec and the Southern Engineering Company in hosting hospitality spaces.

By 1963, Roy Thomas had increased Pike's territory in Georgia to the extent that Glenn Dobson was moved into management. "At the time that Roy asked me to go on the road and help him," Dobson said, "I think that he had already built our work load up to sixteen or eighteen crews. He was traveling not only in Georgia," Dobson continued, "but by then, he had gone into Alabama where we picked up Pike's first work outside our base of operations in Georgia. This was an emergency call from the Central Alabama Electric Cooperative." Contract work followed Pike's quick response, and the company has continued to work there since '62.

In the '60s, many of Pike's first contracts in the Southern Division were with Georgia rural cooperatives whose managers were actual pioneers in rural electrification. For example, early on, Pike obtained work that has continued ever since with the Snapping Shoals EMC at Covington, Georgia, where the co-op's longtime manager was Ed Robinson. Robinson and his family have become life-long friends with all of those at Pike who have known and worked for them.

Rather unexpectedly, Pike obtained work with another of

Georgia's rural electric "pioneers." Bob Armstrong, was the co-op manager at the Central Georgia EMC at Jackson, Georgia. In the early '60s, "Mr. Bob," as Armstrong was called, served as the director of the safety program for the Georgia REAs. When he heard that Pike was working primary, "hot" 12,000 volts, with rubber gloves, he announced that he was going to get "that contractor doing dangerous, hot work 'run out' of the State of Georgia!"

Before that happened, however, Pike was called by one of its own competitor's to complete a pole-changing job for Armstrong's co-op that the competitor was unable to do. Roy Thomas saw an opportunity. He convinced Bob Armstrong to go to the job site and watch Pike's men at work. "When we got there, the crew had already started and was doing a good job, moving the hot primary wires around and working it with rubber gloves."

Thomas said that after they returned to Bob Armstrong's office, the old gentleman softly said: "I need some more of those poles changed out, over there around Monticello, Georgia. Wonder if we could get a crew to do that?" From this auspicious beginning, Pike has served the co-op ever since. "Except for a few times when we've been pulled off to work on storm trouble to help other co-ops," commented Thomas.

Ralph Anderson, assistant vice-president, Central Division, remembered, "In 1960, right after I was promoted to foreman, Roy Thomas wanted me to go to Jackson, Georgia to do a *thirty-day* job with Central Electric Cooperative. I did. And, *nine years* later ...!" Incidentally, Pike general foreman C.H. Long, will retire in 1996, after being at the Central EMC since Ralph Anderson left.

"Mr. Armstrong turned out to be a good friend and one of the best customers that we've ever had," Roy Thomas said. "Later on in the '70s, during a recession when things were really slow there in Georgia, we lost several crews at other co-ops, but 'Mr. Bob' kept our crew working." Thomas said that Bob Armstrong laid off five of his own employees and kept a four-man crew of Pike's working. "That kind of thing makes you know you've really got a friend."

Another of the first Pike contracts in Georgia was with the

Satilla EMC at Alma, Georgia. The co-op's manager, Leman Lanier, is a gentleman greatly respected in the electrical industry, and he and his wife, Merdis, were among the leaders in the early efforts of Georgia's rural electrification. Since 1960, Pike has assisted the co-op in its conversion of 5,000 miles of line. "Without them, I don't know what we would do," said Leman Lanier. "These Pike people who are still here with us now, know our system as well as we know it ourselves." He said that anytime Pike crews are working on his system, they "require the least amount of supervision of anybody" that he had ever worked with.

Pike's rural electrical work expanded into Florida by the late '60s with a contract to Clay Electric at Keystone Heights. Pike obtained work in South Carolina at first with the Aiken Electrical Co-op in '67, and then many other South Carolina REA's became Pike customers.

When the contractor quit, who had been hired to build a substation for the Warren Rural Electric Cooperative in Bowling Green, Kentucky, Pike got a call advising that "if someone gets up there pretty quickly," Pike might get the work. "I called Floyd," Roy Thomas said, "and he jumped in the plane and flew down to Monroe to pick me up. We went directly to Bowling Green and met with Buck Jenkins." Roy Thomas said that shortly after the meeting, a call came from Charlie Stewart, the REA manager there, to say that Pike was hired to start three crews in January '69.

"I had been on that "thirty-day job" in Georgia for nine years," Ralph Anderson said, "when Roy Thomas called and asked if I would go to Bowling Green to start one of those crews."

Anderson began the job with a six- or seven man crew, but within seven or eight months, Pike's personnel serving Warren Electric numbered over fifty. Glenn Dobson and Philip Anderson were also on the job. "We were building transmission lines, substations, and then the distribution lines," Ralph Anderson said. " I stayed up there about a year, maybe a little over a year, when Milton Bumgardener took over for me as foreman, and later, he became a supervisor over that area and oversees the work that continues there today."

*Roy Thomas, (standing), Robert "Mr. Bob" Armstrong, (seated on right with Pike), Opal Pike, (left), and Floyd Pike, (right) seated, center of table.*

*Well, Floyd, I just been thinking about us old linemen.*
*Seems like we'll make good money;*
*but then we don't know a thing about how to save it!*
*I just wondered if you could work out some way*
*for the company to take a little money out of a*
*paycheck each week and save it for us?*
*And when we all retire,*
*maybe then we would have a little nest egg!*

*—Pete Dillon*

# 20

# A More Mature Company

With almost twenty-five years of challenge and success to its credit, "the reputation of Pike's leadership for honesty, dependability, and customer satisfaction was well-established," said Charles A. Simmons, vice-president, American Electric Power Company-Virginia. Pike Electric's experience and growth were manifested in a sense of maturity that was apparent, not only in the caliber of the jobs that were tackled in the'60s, but maturity was evident also in some very important decisions made by management during this time.

New members were added to the board of directors at the stockholders' meeting in 1963. In addition to Floyd Pike, Opal Pike, and Emmett Forrest, Jr., nominations were presented by Thomas Scott to elect the following: Joe B. Pike, Z.W. Blackmon, Sr., A. Haywood Merritt, Jr., and Roy W. Thomas. Ninety-nine stockholders were either present or represented at the annual meeting, and the motion was unanimously carried.

Two years later, a motion to adopt a deferred profit-sharing plan was presented by A.L. Mc Alexander at the annual August meeting. The idea for a profit-sharing plan is attributed to Pike foreman, Pete Dillon. Dillon's dedication to the company and to his fellow employees began in 1945 when Floyd Pike gave him his boots in order to work. Committed to the company's success, he was among the first employees to purchase stock in the company, and through his candid suggestion for planned savings, Pete Dillon influenced and contributed to the company's future. In less than a year, procedures were in place, and the board of directors unanimously voted to establish the "Profit Sharing Retirement Plan" in June 1966.

Original endowment for the fund was obtained through purchase and contribution of Pike Electric Common Stock with value of $200,000. "From its beginning, it was not a forced, or even an optional, plan," Joe Pike said. "All the funds come from the company — out of its profits. No matching or withholding has ever been involved."

The plan grew rapidly and exceeded half a million dollars during the next two years, as the company provided continuing contributions to the plan through additional stock purchases. Since it began, the retirement Plan has been modified through the years to meet government requirements (ESOP) and to increase personal benefits to become a profit sharing, plus 401(K). Today it is approaching thirty million dollars in value and has paid out over eleven million dollars in retirement benefits in the last ten years.

## Corporate Financing

Late in the 1960s, further expansion — and how to accomplish it — produced the greatest financial challenge the company had ever had known. More capital was essential.

Through the company's association with Edwin Duncan, Sr., and the Northwestern Bank, Floyd Pike was introduced to the corporate investment personnel of the Metropolitan Life Insurance Company and directed to Robert Schwartz, MetLife's former chairman, president, and chief executive officer. "My first contact with Floyd S. Pike Electrical Contractor, Inc., was in early 1969," Schwartz recalled, "when Floyd approached Metropolitan through Northwestern Bank in North Wilkesboro regarding our interest in a loan."

Although the amount of money that the company needed at the time seemed huge to Pike's management, "that size transaction was unusual for Metropolitan," Schwartz said. "Our average loan amounts, even back then, were considerably higher, and there were those here who didn't want to get involved in such a small financing."

Despite any doubts of others at MetLife, Robert Schwartz cleared the way for the loan to be made. "I had a good feeling about Pike and Floyd's 'can-do' attitude. He had taken a concept, developed it into a working organization, surrounded himself with good people, and turned that $5,000 investment of his into the going concern Pike had become by the time MetLife made its loan." In 1969, MetLife loaned Pike Electric, three million dollars — definitely, another point in the company's maturity — and this was only the beginning of an association that proved worthy to all parties.

Essentially, the loan decision was based on the company's integrity and promise for the future. Robert Schwartz observed:

> It's sometimes difficult in a large organization like MetLife to see how our individual actions affect, either positively or negatively, others in the organization, and similarly, how our actions can influence customer decisions. But it was easy to see how it all came together at Pike.

> One of the strongest memories I have of Floyd and the other officers was the attention they gave both to employees and customers, and the compassion and understanding they expressed for their needs. It was clear to me from day one of my association, that respect for employees and their contribution, fair treatment of customers in business dealings, and encouragement of a team spirit in conducting business were playing a vital role in the company's success.

Incidentally, Joe Pike said that when the transactions for the initial MetLife-loan were completed that day in New York, "Dad and I left MetLife and went directly to Chase Manhattan and deposited the check. We didn't waste one minute, putting that money to work!"

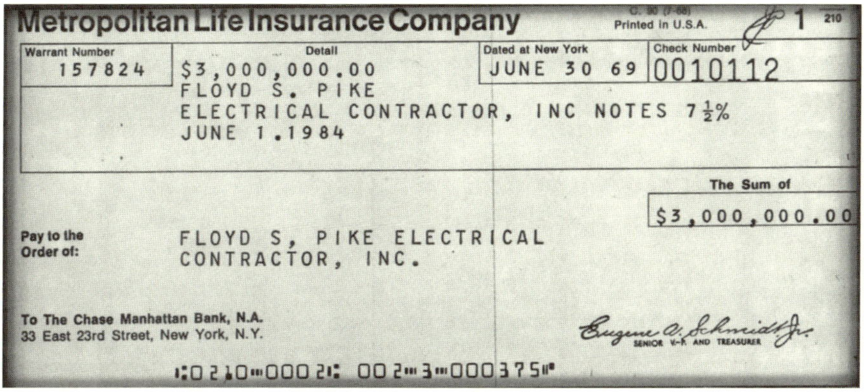

Metropolitan Life Insurance Company

C. 90 (7-68)
Printed in U.S.A.

1 210

| Warrant Number | Detail | Dated at New York | Check Number |
|---|---|---|---|
| 157824 | $3,000,000.00<br>FLOYD S. PIKE<br>ELECTRICAL CONTRACTOR, INC NOTES 7½%<br>JUNE 1.1984 | JUNE 30 69 | 0010112 |

The Sum of
$3,000,000.00

Pay to the Order of: FLOYD S, PIKE ELECTRICAL CONTRACTOR, INC.

To The Chase Manhattan Bank, N.A.
33 East 23rd Street, New York, N.Y.

Eugene A. Schmidt Jr.
SENIOR V-P AND TREASURER

⑆0210⑈0002⑆ 002⑈3⑈000375⑈

*"We didn't waste a minute putting that money to work!"*
*Joe B. Pike*

# Up and Down and Up and Up and Up and Up

*Back in the early '70's,*
*our heating and air conditioning people*
*were installing a new heat pump for the Pikes at their home.*
*Mr. Pike asked about the problems they were encountering,*
*which were caused by the retaining wall.*
*I began to explain the difficulties to him by saying,*
*"You can't pull all those elbows in the refrigerant lines without causing —"*
*Before I could finish my explanation,*
*Floyd quickly turned to me and said*
*"Wait just a minute, young man!*
*We got rid of* can't *in this company in 1945,*
*and I don't ever want to hear you say it again!"*
*— and he walked out.*
*I had been with the company one week!*
*—Butch Riley*

# 21

# Climbing In The '70s

With the corporate financing secured, the decade of the '70s opened with optimism. Pike's work on the nation's Southeast power grid continued with huge transmission jobs. Contract work began for new customers, Georgia Power Company and Florida Power Corporation. And new jobs developed in Ohio.

"In '73, we were awarded the distribution contract with the Portsmouth division of Ohio Power Company," said Richard Simmons, supervisor Central Division. "And the company has worked for Ohio Power Company ever since."

During the same period, Pike began working for the rural electrical cooperatives in Ohio as well. The first job began in the only too familiar context of a strike situation when Pike crews were called to assist the Buckeye Rural EMC at Gallipolis, Ohio. Working for a new customer in an emergency was tense enough — with the usual union harassment — but conditions became even more stressful when the area was hit by an ice storm. Despite the difficult beginning, Sammy Hall and his crew did an exemplary job that resulted in a growing and continuing association with the Ohio electrical cooperatives.

In North Carolina, Tom Scott was converting overhead lines to underground in Charlotte's "Myers Park," with the help of Zack Blackmon, Jr. Blackmon, together with "Smiley" Phillips, Billy Martin, Herb Angel, Haywood Williams, Frankie Riddle, Albert Ruff, Kenneth Caudill, Bobby Riggs, Lowell Anderson, Jerry Chinault and Bobby Sapp moved on to work for Lawrence Allen.

Allen was overseeing the construction of several substations, and he was in charge of converting the overhead distribution system from a 4,160 to a 12,000-volt system on the military base at Fort Knox, Kentucky. He also directed the re-wiring of the Radford Army Ammunition Plant, the arsenal in Radford, Virginia. And during the early '70s, he was in charge of crews that were involved in the total wiring, including cable and telephone, for several new suburban residential communities, like Big Canoe outside Atlanta, that were then under construction.

In addition to the numerous, big jobs that involved many crews and tremendous amounts of equipment, the gamut of the company's work in the early '70s also covered an unusual job in tiny Tate City, Georgia.

It seems almost inconceivable that as recently as twenty-five years ago there could exist any town in the United States still without electricity. America had put a man on the moon in 1969! Nevertheless, for a job that seemed like an anachronism, the Pike company was summoned to Tate City, when then governor Lester Maddox, issued an order that the residents of this tiny town in the north Georgia mountains, would no longer be without power.

"From the take-off point at Titus, Georgia, near Hiawassee, the line was constructed overhead," said Southern Division area supervisor Gollie Easter, who was then foreman on the job, "but it was put underground at the Appalachian Trail, and continued again overhead across the Blue Ridge Mountains." He said that the line required the setting of 75 poles in terrain so rough that crew members Clyde Nichols and Grover Smith had to dig most of the holes by hand. Gerald Nelson and Stanley Berryl helped set the poles by using "piking poles, dead men, and an old military-type '69 Dodge power-wagon, that required chains on all four wheels to get up and down the mountains."

"Working in such difficult conditions," Gollie Easter said, "the mayor of Tate City, Arthur Young, tried to make things easier for us." He already had the right-of-way cleared for the crews before the job began, and he wanted to help with setting the poles by supplying a bulldozer. "But over the weekend before we began, someone stole the starter from the 'dozer'!" explained Easter.

After Pike completed the job the the Blue Ridge EMC, television crews arrived from Atlanta to film the "progress" that had taken place, since Tate City was the last town in Georgia to receive electricity. "The people of Tate City," said Gollie Easter, "were very happy when the lights came on!" He concluded, however, that "this was one tough job for one tough crew!"

## PIKE WIRES IN COSTA RICA

Pike soon carried electrification into a more surprising rural destination. In the early part of the decade, a partnership in a Costa Rican banana plantation resulted from the friendship Floyd Pike shared with Edwin Duncan, Sr. — a friendship that had grown even stronger by '71, when Duncan consented to Pike's request to serve on the company's board of directors. Their Costa Rican partnership included several other North Carolina businessmen, all of whom had worked together since the early '60s in their joint efforts with the Northwestern North Carolina Area Development Association. "The banana plantation, near Siquirres, east of San Jose, Costa Rica, was the setting for many good times," Opal Pike said, "but for Floyd — to separate work and leisure was never possible or desirable, even at that distance!"

Adjoining the plantation were small villages where the farm-laborers lived. Floyd Pike was both aware and disturbed by the fact that one of the villages lacked electricity. He knew that he could easily remedy that situation and that doing so would provide the workers and their families a much better standard of living. Floyd Pike asked Leo Sams, Pike Electric supervisor, Northern Division, to go with him to Costa Rica and build a connecting power line.

"One of the villages — that included a few houses and the school — had a fairly reliable power supply," Sams explained. "A little turbine generator worked from a small hydroelectric plant that was down the mountain on the Revantazôn River." The adjoining village had about fifteen or twenty houses, along with the movie theater, but the only power supply for that village was a diesel generator, with a motor that was "going bad."

Leo Sams said that when he and Floyd Pike arrived in San Jose, Sams was directed to go immediately to the streets and deal with the money-changers in order to have the payroll when the work was completed. "I quickly and correctly, determined that this was not just a job, but a real adventure!"

Floyd Pike and Leo Sams flew in a small plane from San Jose to the farm where the work was already underway. Sams reported that the farm workers had "gone into the woods, cut down trees, and made poles, 'saw-logs,' that were 25 or 30-feet long, and, sometimes, 2-feet across at the top." The workers had set and tamped the poles — "all fifty-three of them" — and they had mounted the cross-arms on some of them.

"I told Mr. Pike that I would need ten people to finish framing the poles and building the line," said Leo Sams. "And then he told me that my 'crew' spoke only Spanish, and my interpreter spoke only a little English." "Oh, Boy! This is going to be fun!" Sams said to himself. "And it was!"

He directed the interpreter to work as "the foreman," and together, they got production moving by dividing the men into three crews — some continued the framing and others began pulling wire. "I had the only climbing gear," Sams said. "The workers had built their own ladders to reach the top of the poles."

Lacking materials in their remote surroundings, the workers had built cross-arms from hard wood (4"x4"x4') and cross-arm braces from galvanized water pipes. They did not have the "bits" usually used in mounting the cross-arms, but they substituted with U-bolts, made from steel rods, placed around the pole and through the cross and, thus, they accomplished the task. After all the pole work was done, Sams said that the workers tired quickly, when pulling the wire. "When they could pull it no further, we simply put the reel of wire on the farm tractor and drove it on down the road."

Leo Sams said that they "built the line, made all the connections, installed six lightning arresters, and had all lines feeding from the turbine power plant in about twenty-four hours." Then he and Floyd Pike caught a train for San Jose. "It was a great trip!"

*Leo "Whity" Sams and "crew" in Coast Rica.*

*Floyd Pike and Zack Blackmon in Costa Rica.*

*To reach banana planation, one travelled by plane, train or horseback.*

*Costa Rica Village*

*Poles were made from "saw-logs."*

*Costa Rican Village*

## DIVERSE JOBS

At the same time, Leo Sams was building lines with men on handmade ladders in the Costa Rican rain forest, Henry Coe was supervising a high-voltage transmission line for one of North Carolina's mountain-top resorts, "Wolfe-Laurel," in territory so mountainous that the job could only be done by helicopter.

"And then there was the Potomac Edison Power job outside of Washington, that river crossing," said Jasper Marion, who came to work with Pike in '57 and who has been in charge of some of Pike's most ambitious projects. "We were to build four steel towers, one on each side of the river, and two in the water, for a transmission line to cross the Potomac River."

The job itself was controversial on two accounts: first, permits had been issued for a line to be built in "anticipation" of future electrical needs. And, secondly, the location of one of the towers was causing dispute. Actually, the tower's location was legal, that is, it was located outside the shipping lanes, but it did stand in an area of the river that ships actually favored. "Crossing the Potomac was altogether different from anything I had ever done before," Jasper Marion said. "We used tug boats routinely, because we had to do all of our work from barges. We had a small yellow boat that we used to ferry ourselves out to the job site. Every time we left it and a ship would pass, or the current was rough, the boat would be swamped. We called it our 'yellow submarine!' "

Shortly after the Potomac River job, Jasper Marion moved on land to supervise a steel-tower construction project in Charlotte for the Duke Power system. The equipment requirements on this job contrasted greatly from those needed previously. "On the Charlotte job, we actually used a horse to pull the spacer buggy," Marion said. "It was even on the payroll!"

*Improvement or betterment for our people is our own job.*
*— Floyd S. Pike*

# 22

# Pike's Support Services Deal with a Fast Pace

Accounts of some of the jobs of the '70s indicate not only the tremendous growth that was occurring, but they also point out the escalating demands in mobility and distance that were required of Pike's personnel.

"Very few employees have ever really known an eight hour day," said Donald Anderson, supervisor, Central Division. "In the early '70s, I was transferred to a job in North Wilkesboro that the company was doing for Duke Power, a job that that lasted for five years. In addition to working an eight-hour day, Kenny Anderson, my foreman, and I traveled at least two or three hours a day to get to work and back." Donald Anderson added that the drive was on "old two-lane roads, five days a week." And this was typical.

"A little later, I worked with the garage and the fleet," Anderson said. "Then I was able to see everywhere we worked and meet a lot of different people that worked for the company. Early in my career, I actually saw how fast the company was growing in the early '70s. I really got a feeling of how big our company was, how spread out it was, and where it was going at that time."

The expansion that the company was experiencing was pressing hard on all of the support services within the company. In the garage, A.L. McAlexander and his foreman, Robert Johnson, had their mechanics and other garage personnel working two shifts a day to keep the equipment operating. Don Edwards somehow managed to keep up with the parts that the garage and the crews needed from a tiny office in the middle of it all.

"Low-boys" — trucks that are capable of hauling heavy equipment — were not yet available in the early '70s, so additional wear occurred when a piece of equipment was driven, rather than carried, to the job site. Traveling on roads that were badly damaged by storms also added wear and tear to Pike's equipment, not to mention the great distance that the company's work now covered. Witchia Gravitte, transportation supervisor, said that sometimes when the runs were long, he and his brother, Hylton Gravitte, and Ralph Puckett, would be on the road for two-weeks at a time. The trip to Louisiana was two thousand miles, and often, the drivers would leave there and go directly to Maryland. "We'd just come by the house long enough to say 'Hello' to the wife and kids, and then go back to Louisiana. There were some pretty hard times back then."

The company's fast pace required additional assistance in the main office. In 1971, Jean Williamson, executive secretary in the corporate headquarters, came to work with Pike, a company she knew nothing about at the time. Having just moved to Mount Airy, she said that when Haywood Merritt called her to come the office for an interview, his directions for finding the place ended with: "You'll see little yellow trucks dotting the hillside!"

Jean Williamson said she thought the office might be "just a little hole in the wall, with just an employee or two. "I was surprised to see an array of yellow trucks, the size of the building, and so many employees!" Williamson said she quickly became aware of how widespread Pike's operation was when three of her five bosses were away at the time that she was hired. "Mr. Pike was in Costa Rica," she said. "I don't know where Zack Blackmon, Sr., was. And Lawrence Allen was working at Fort Knox. For many months, before I ever met him, he'd come in on the weekends and leave work for me to do. He was like a phantom boss!"

## EMPHASIS ON SAFETY

As Pike Electric expanded and personnel numbers continued to increase, additional supervision and consistency in the emphasis and requirements of the company's safety procedures became necessary. Added to these considerations were more government-mandated safety regulations issued by the Occupational Safety and Health Administration. [OSHA]

Floyd Pike reviewed the qualifications essential to the job of upgrading and systematizing the safety department. A.L. McAlexander was the man for the job:

> I had a background that related to safety work. During World War II, my Army occupational specialty was No. 1 - explosive technician, as well as No. 1 - motor vehicle repair mechanic. I had been a volunteer fire fighter for almost ten years. I was a First Aid instructor and a state-certified instructor of fire fighters in both Virginia and North Carolina. I had been exposed to quite

> a lot of power-line work and terminology, and I was an instructor of radiological monitors under the old Civil Defense program. Also, I felt I was persuasive enough to talk people into working safely.

Without hesitation, Pike turned the responsibility of the safety department over to McAlexander. Dwight McAlexander carried on the garage operation for his father.

At the time, the company established its own purchasing arm, Pike Electrical Equipment and Supply. This subsidiary provided the company with ready access to often used supplies, which included "Kine-pak" explosive products. A great deal of the company's underground work was, and is, located in city-centers causing safety education and training in explosives to assume new and critical importance — far different from the days of dropping a few sticks of dynamite in a hole as it had been done in the '40s and the '50s. McAlexander s qualifications were ideal!

Changes in job requirements and equipment developments made the company's safety manual obsolete. Along with writing a new handbook of rules and devising new methods for teaching job safety, McAlexander was frequently on the road meeting with crews to implement and enforce new safety directives. Zack Blackmon, Jr., safety director, following A.L. McAlexander's retirement in '87, said: "I had the opportunity to work with 'Mack' before he retired. He could be labeled by all who knew him as honest and precise. He was a person who, in essence, knew something about everything. You'd better have some time to listen if you ever asked Mack a question. He was thorough!"

A.L. McAlexander's personal characteristics were the very qualities that were essential to accident investigation — an integral aspect of his of duties during the fifteen years he was safety director. He had been on the job in his new assignment for only a few weeks when his investigative ability was acutely tested as a result of company policy.

The hostility that Pike's crews had confronted when the company began work for the Kentucky Power Company, reached a new level of severity in the early '70s. A.L.

McAlexander was dispatched to Ashland, Kentucky, in 1972 to investigate the company's first instance of industrial sabotage. The explosion and destruction of a crane being used in transmission work there only served to reinforce the Pike company's non-union policy.

**Ashland Daily Independent**

☆ ☆ Associated Press Leased Wire and Wirephoto Network 20 Pages Today

# $25,000 Reward Is Offered In Power Crane Destruction

By DAVID CRANCE
Staff Writer

Kentucky Power Co. is offering a $25,000 reward for the arrest and conviction of those responsible for the explosion here Wednesday evening of a crane being used to construct a transmission tower.

The crane belonged to Floyd S. Pike Electrical Contractors, Inc., which is under contract with the power company to erect the tower.

Local, state, and federal authorities are investigating what Jack Palmer, public affairs director for Kentucky Power, has called "industrial sabotage."

In announcing the reward, Mr. Palmer said, "The public is advised that information on such unlawful acts can be furnished to the company or to any law enforcement agency of the state or federal government or to the sheriff of any county or to the police of any municipality. The reward is also offered for any person "aiding and procuring any person to do such an act."

The reward offer expires June 1, unless extended by the company, and "the total amounts of the rewards payable for all instances shall not exceed $100,000."

Mr. Palmer said the act of sabotage "endangers the public dependent upon electricity for safety, health, and comfort."

He also said that construction of the tower will continue under the contract with the Pike company.

An official of the U.S. Treasury Department, investigating the incident, said "two charges, possibly dynamite, were used to destroy the crane with one charge larger than the other. One charge was placed on the hoist control motor of the crane and the second charge was placed in the driver's cab area."

He also said the charges were set to explode simultaneously. Plaster cast footprints and measurements have been taken at the scene of the explosion and the evidence has been sent to treasury laboratories in Washington, D.C.

The spokesman said the footprints around the scene showed the possibility of three people being involved in the sabotage and that the tracks came and left the scene from the west.

He has requested that anyone living in the area who saw or heard anything unusual between 7 and 9 p.m. Wednesday to contact Kentucky State Police at 928-6421 or call the Treasury Department at 324-6677.

The agent also said the saboteurs had apparently used a burning fuse and had to run quickly from the area to avoid injury from the explosion. The Pike firm has its main offices in Mount Airy, N.C.

★ ★ ★ ★ ★

$100,000 CRANE DEMOLISHED—This crane belonging to Floyd S. Pike Electrical Contractors, Inc., was destroyed Wednesday night when two high explosive charges were detonated on the vehicle. The firm is under contract with the Kentucky Power Co. to erect a transmission tower located at the west end of Armco Steel Corp. property. No one was injured in the blast, although the explosion was heard for several miles around. (Staff Photo by Crance)

## PIKE IS A NON-UNION COMPANY

From its beginning, Pike Electric has always been a non-union company. Union policies and tactics have no place in the professionalism of Pike Electric's work force. The company's history shows that, all too often, Pike crews have endured significant loss in their own productivity and accomplishment, while dealing with the harassment and indignant behavior of union workers whose conduct has sometimes even jeopardized the personal safety of Pike employees.

"Our non-union policy goes back to the company's beginning," Joe Pike explained. "Dad declared that this company would never be a union company." Today, with ownership shared by management and employees, the Pike company will remain non-union.

"We are not only a part of the company," said Robert Chilton, safety supervisor, "we are owners of this company, as well." Employee ownership places the company's accomplishments into the hands of the individuals involved. "This gives employees the realization that they have helped to perform what has been achieved," said Jim Draughn, supply supervisor. "When we look back and say, 'Oh, Boy! We've really been successful!' We can also say, 'I was a part of that!'"

The qualities of independence and respect for the individual, his opinion, and his ability, are very important among all Pike employees. "We, as employees, have never had to have someone else speak for us," Robert Chilton said. Indeed, it is Pike's open-door policy that is considered to be its key to good working relationships. "Any employee can talk about a problem to anybody in management that he wants to," said Ralph Anderson. "He can talk to his foreman, his supervisor, or he can go to Joe Pike, if he's got a problem." And one can count on an honest response.

Ralph Anderson said that Pike is always "up front" with each of the company's employees, whether the person has been with the company "two weeks or forty years." An honest response is an integral part of Pike's integrity — and the very foundation of the company's reputation. "The most important thing about this company is its integrity," said Akers Woods, general foreman, Pike garage operations. "It has always been important to me to say that I work at Pike, for the qualities and values that this company represents." "The standards are the same for everyone that we deal with," Ralph Anderson added, "— employees, customers, and suppliers, alike."

Jim Draughn' s opinion is that Pike, with its non-union position, has set an example for fifty years. "Pike is a place where employees can work for a non-union company and still be treated fairly and earn an honest day's pay for an honest day's work. I feel that is also a part of this company's integrity. This is something that makes the company strong."

In discussing their history of working in a non-union environment, many Pike employees referred to efficiency — in the field and in management — as a practical advantage of the company's position. "We want to get the job done!" Witchia Gravitte emphasized. "For example," he said, "if a truck driver wants to adjust his own brakes on his own trailer, he can do it without somebody telling him that the company's mechanic has to do it." Robert Chilton agreed, "We have always been able to do any part of the job that we were sent out to do, without being told that we could pick up this or pick up that. This way, we have been able to learn faster and do more. You can do whatever you are qualified to do here."

Efficiency in management requires flexibility, rather than rigidity, to meet the spontaneous demands of a company's day to day operations; and, moreover, to amend a long term plan, when it is advisable. Barny Ratliff, executive vice-president, said that he had worked in a union environment and had experienced only "frustration." He said that any personnel decision, regardless of how insignificant, had to be "referred to the contract." Ratliff added that the worst thing that a manager could do in a unionized setting was "to err in the contract's interruption" and "set a precedent." Then a company might then be obligated to a course of action that was actually harmful to its business. Ratliff cautioned that "every imaginable situation cannot be rigidly defined by a labor contract, and simply trying to apply an interpretation that most closely fits an issue can create new difficulties."

Does working with many of one's own family members promote or insure good working relationships? Or do good working relationships at work encourage family members to join others in their place of employment? Whatever the answers, Pike employs hundreds of persons whose family members also work with Pike. Many are second- and third-generation employees. "Just look at all of the father, sons, and daughters working here," said Ralph Anderson. "My older son, Tony, has been with Pike for almost sixteen years; and my younger son, Chris, has been here a little over twelve years." Evidence of good working relationships speaks for itself.

Jerry Boatright, assistant vice-president, Southern division, said that the "family environment" is strongly felt within the Pike company. "In a work environment," Boatright said, "all employees should work together for the betterment of the company and be as one big family." The family environment indicates a unity in goals and caring that is widely apparant among Pike personnel. "This goes to show that we have a company in which people who are willing to learn, and to work, can certainly advance and, at the same time, be happy at their jobs," said Boatright. "In my opinion, a 'union' only divides and separates," he added. Barny Ratliff concurred: "If an 'us-against-them' attitude prevails, the situation then becomes one that is totally counterproductive."

Pike's history is filled with accounts that demonstrate how important the company's non-union position is to its customers. How many times have Pike's crews provided assistance during strike situations? "We would not be working for some of our long-time customers now, if we were union contractors," stated Jerry Boatright. When Pike crews are hired, the customer is assured that a strike will not interfere with getting a job completed.

This policy is reassuring to Pike's own personnel as well. "I've been able to work throughout the 40 years that I've been with Pike without being on strike," said Robert Chilton. "In my traveling around through the country, I've seen people on strike over a year; three months; six months — stuff like that. Belonging to a union, you never know when you're going to be able to work, and when you're not going to be able to work due to someone else's disagreement." Chilton concluded, "It's been a blessing to me to work for a non-union company."

"Robert Chilton is right!" said Joe Pike, commenting on union procedures. "A real benefit to us, and to our customers, has been that any work stoppage or slowdown in the history of the Pike company has always been the result of an industry slowdown — and not from internal problems or disagreement. This really gives our customers additional surety that when we say that we will get the job done, we will!"

Joe Pike said that, when he and Floyd Pike, along with Haywood Merritt, went to Kentucky to join A.L. McAlexander in the investigation of the crane-demolition, "The situation there only proved that the Pike company wanted no part in those kinds of union disruptions." Pike continued, "If there are problems or dissatisfaction about something, management wants to know about it immediately." He emphasized that "anyone who comes to work here learns very quickly that we are up-front with everyone, and with Pike's foremen, our supervisors, right on through the management team, there is always someone available to help with a problem or to correct a difficult situation."

Fundamentally, Pike Electric's policy is simply a commitment to taking care of itself. "Our policy has always been based on our belief that all of us working together can solve our own problems," Joe Pike said. "We provide for our own benefits and for our individual retirements with our own profits. The profits we earn honestly, through hard work and good management."

Joe Pike added that, from the beginning, Pike Electric has worked not only to succeed, but also to share that success with everybody involved here. "That's the way it has always been. That's the way it will always be. That's the way we operate."

"'Improvement or betterment for our people is our own job,'' Joe Pike said, quoting the words of his father, "whether the 'improvement or betterment' applies to safety, individual benefits, equipment, or anything else we do."

*To "go public" required a lengthy audit–*
*so lengthy that before the results were completed,*
*the stock market had taken a real turn for the worse*
*and everyone advised us against going public,*
*at least for a while.*
*By the time the market took a viable upswing,*
*we had solved our long-term financing,*
*and the idea of a public offering was never again considered*
*In my opinion, this was the pivotal point in Pike Electric's history.*
*—Joe B. Pike*

# 23

# Going Public

On the wall, for all to see, was evidence of an ever-widening perimeter of the company's service area. By the early '70s, the wall map — dotted with its multicolored pins — was no longer adequate to visually track Pike's extensive operations. While computers were still in development, new mapping boards that covered an entire wall, sixteen feet in length, were installed in the main office in Mount Airy. Roger Moody, license supervisor, now retired, weekly updated the organizational chart. Color-coded tiles designated the location of all company personnel, the numbers and locations of all crews and the various pieces of equipment on each job.

The growth indicated by the map had not come without tremendous efforts and costs. The growing pains continued to be felt. New equipment was now available and more equipment was necessary. More people were needed, and in more areas. With all indications justifying confidence in the company's record of achievement and its bright prospects for the future, more capital was crucial to meet Pike's needs and facilitate continuing expansion.

In 1971, Floyd Pike called a special board of directors' meeting in the company's headquarters in Mount Airy. To obtain the necessary capital, he presented the company's options as he saw them: a public stock offering; a merger; or a loan — substantial in size and structured for the company's projections. All options were carefully and thoroughly explored, and a public stock offering seemed the way to go.

"Floyd Pike had made some initial contacts with investment bankers," said Jim Smith, an audit partner with Ernst & Ernst (E&E, now Ernst & Young), "and he had been informed that the company should use a national CPA firm in an initial public offering of capital stock." Smith said that he and his partner William Shelton, came to Mount Airy and met with Floyd Pike, Opal Pike and Emmett Forrest to discuss what the "auditor's involvement" was when a privately owned company decided to "go public," and E&E was chosen to carry out the accounting.

But "going public" was not to be. Joe Pike explained that "procedures came to a stand still for what seemed like all the wrong reasons, the audit seemed endless. It was hard for anyone to realize at the time that the events were actually a blessing in disguise." When the market took its dive, a new decision was made — to obtain a loan, ample in size and structured long-term, to carry the company on through a presumably profitable decade.

Minutes from the meeting record: "The Chairman, Floyd S. Pike, stated the corporation was needing more capital to grow. After discussion, the motion was made by Edwin Duncan, Sr., to authorize Floyd Pike and Joe B. Pike to negotiate and consummate a loan of an additional two million dollars with Metropolitan Life Insurance Company in New York." Once again, Pike contacted Robert Schwartz.

"From MetLife, not only was the company provided the substantive money it needed, but the personal relationship that had begun with Bob Schwartz, and that continued with Bob Chappell, provided professional advice and assistance that was extraordinarily helpful," said Joe Pike. "Mr. Schwartz served on our Board from '72 through '75 and, in my opinion, the perspective of the entire management team was changed." Joe Pike added that at every meeting, Robert Schwartz not only contributed with his understanding, but he always shared with the board the benefits of his knowledge and his experience. "While we kept our focus," Pike said, "Robert Schwartz caused us to look at our actions, and our direction, with new dimensions of possibility."

On a personal note, Pike added, "I certainly regard Robert Schwartz to be a true friend. He was present at the board meeting held in Mother's kitchen after Dad's death, when I was elected president. Robert Schwartz never failed to offer valuable guidance and advice any time that I called. He stood by me — and he stood by the company — when the water got pretty deep! It was always a pleasure to have him, and to have Caroline, in our home."

Another important fact of Pike's history should not be forgotten. It had been the guidance provided by Edwin Duncan, Sr., that led Pike Electric to MetLife in the first place. The association that the company had shared with Duncan, for twenty years, sadly came to an end when he died in 1973. A tribute from the company minutes reads:

> Upon the death of Edwin Duncan, Sr., on October 7, 1973, this Corporation, along with the people of northwestern North Carolina, lost a true and valuable friend. His guiding hand has been instrumental in this company's growth. His quiet advice, humbly given, always seemed to put a problem in perspective. His knowledge and experience were great assets in determining our course of action. Since his election to the Board of Directors, Edwin Duncan's contribution to this company increased even more. He did

not miss any of our scheduled meetings. He was a real inspiration to the other members of the Board. He will be sorely missed.

*Stockholders' meeting in the early '70s. Robert Schwartz, left.*

*I will give Reg Banner credit, to a large degree,*
*for our becoming aware of which jobs were profitable,*
*which jobs were not profitable,*
*and where we should expend our energies.*
*Reg could determine what it cost us*
*to run any kind of equipment for an hour*
*better than anyone I've ever known.*

*—Haywood Merritt*

# 24

# Euphoria in the Early '70s

Another benefit, which resulted from the seemingly interminable audit that preceded the MetLife loan, was the company's employment of one of the young accountants sent to Pike from Ernst & Ernst. "After a couple of years on the audit, Emmett Forrest asked if I wanted to come to work for Pike Electric," said Reg Banner. "That scared the hell out of me! I was thirty years old and recently out of college; and, for the first time in my life, I was making more money then than anyone else in my family ever had!"

He said that his interview with Floyd Pike, Joe Pike and Emmett Forest was an unforgettable experience. "After I promised to work harder than any other accountant on the face of the earth, Floyd looked at me and said, 'Boy, we ain't hiring ya,' we're marrying ya'!' " Reg Banner declared, "If ever an accountant landed a dream job, I did!" He added, however, "When I look back now at the numbers, the company wasn't very big then, when compared to today; and there were no guarantees or assurances that it would flourish."

Reg Banner may have taken a chance when he came to Pike; but he allowed no chances to be taken with Pike's numbers. A certified public accountant, with a background that included work experience with a "big eight" accounting firm, Banner quickly went about establishing procedures to accurately determine the company's cost of doing business. "Pike was in desperate need of updating their accounting procedures and policies," Banner declared. "Their monthly financial statements consisted of two pre-printed pages produced on the old multi-lith copiers." He explained that one page was the balance sheet, and the other was the income statement. Monthly statements were prepared by inserting the numbers in designated areas. "Using blue ink, we would simply write in the numbers," he said. That method had been used for 27 years.

Unlike in some situations, where change is met with resistance, Reg Banner said, "I was appreciated and accepted as a part of the team from the start. Those early days were fun and exciting." He added, "We all worked together to make things better."

With work at an all time high, Reg Banner could not have joined the Pike team at a better time in its history. His records indicate that from 1969 — the year of the initial MetLife loan, until '74 — the company's annual revenue almost doubled. "Floyd knew we were close to making a million," said Banner, "and he was so hopeful that we would not find any downward adjustments that would bring it below the magic figure. I knew we wouldn't, if I could help it." The company did make it — "with a little to spare." To Banner, that achievement appeared to be one of Floyd Pike's proudest moments.

*Reg Banner.*

*The '40s, '50s and '60s were decades*
*of enormous growth for us. During those years,*
*the company established "partnerships"*
*— in the finest since of the word —*
*with numbers of our customers.*
*Later on, in the '70s, when our industry was hit,*
*at least as hard as any, by Jimmy Carter's "malaise,"*
*and other factors beyond our control,*
*we had customers who were forced to*
*cut Pike's forces drastically.*
*But many of our customers then resorted*
*to cutting their own crews,*
*rather than dismiss us completely from their systems.*
*Others took innovative steps, such as*
*making some of our crews "multi-functional."*
*Duke Power Company, for example,*
*had some of our crews working*
*as line crews for two or three days a week,*
*and the rest of the week as tree-trimming crews.*
*These helpful measures were the result*
*of some very strong working relationships*
*that were established long before*
*"partnership" became a procedural trend.*
*I will never forget the loyalty of these customers.*
*The loyalty of customers like these became*
*crucial to our company in the '70s.*
*The loyalty of these customers and the individuals*
*who often went "out on a limb" for us then,*
*and who have continued to hire us,*
*will never ever be forgotten by the Pike company.*
*—Joe B. Pike*

25

# A Storm of Our Own

In 1974, the hard work of Pike's personnel culminated in the company's achieving its most successful year to date. However, the time for savoring success was short-lived. Company minutes from the annual shareholders' meeting that year indicate that a review of the company's accomplishments was accompanied by an ominous prognostication: "The Chairman, Floyd Pike, stated that the business outlook for the coming year looked uncertain, due to the economy and the problems facing the utility companies." At that point, no one could foresee the serious consequences of this prediction.

In *Duke Power Company; The First 75 Years,* Carl Horn, who was then president of Duke Power Company and in '75, became Chairman, explained:

> The most critical, frustrating time we ever faced at Duke Power was the 1973-'74-'75 period. There was the recession and the Arab oil embargo to deal with during that time. Our bond rating dropped in about a four-year period from triple-A to single-A, and there was a period in 1974 when you couldn't sell a long term single-A utility bond at any price .... Compounding all of that was the negotiations going on with the United Mine Workers of America .... To further compound our problems in that period, we had to ask for a rate increase .... Holders of large construction contracts had to send messengers down here by airplanes to pick up their checks. At the same time, the UMW picketed the New York Stock Exchange and placed full page ads in the *Wall Street Journal* trying to get people to refuse to buy our stock.... [16]

The situation was just as grim for other long-time customers. Charles A. Simmons, then a young executive with AEP's Appalachian Power Company, explained that the early '70s marked "the beginning of a difficult era in the electric utility industry" that was caused by "rising interest rates," and "sharp increases in fuel costs." In addition, he said that "environmental expenditures to comply with the Clean Air Act, plus a slowing of the economy, all combined to reverse the trend of decreasing electric rates." Simmons explained further that "regulators, who had previously found it relatively easy to decrease rates as costs fell, were unable to come to grips with the need to increase rates, as costs increased." He said that the consequences of all of these factors resulted in the inability of the utilities to increase rates. "Our first rate case took nearly five years to complete with little relief granted in the end. Costs, however, continued to soar, which then led to bond deratings, and interest coverage ratios that threatened to precluded further borrowings, and plummeting stock prices. "In short," said Simmons, "APCO was facing a cash flow crises which required drastic cuts in expenses."

The ramifications of the industry's dilemmas created a major impact that shocked the Pike company. Appalachian Power Company, like many other utilities, made the "the difficult decision to release all contractor forces, including Pike, in late 1974 and early 1975," said Simmons. He observed that this was definitely "a black time in an industry that had never known layoffs previously."

Reg Banner said that Pike's employment dropped from approximately 1,500 to 800, almost immediately. "We were getting calls daily to release crews as the utilities reacted to the paranoia of Wall Street." Haywood Merritt said that after all of the crews serving the Appalachian system were released, Floyd Pike personally appealed to John Vaughan, then the president of Appalachian Power Company, and "even in those difficult times, John Vaughan, kept a Pike crew working."

Stemming from world wide events and the electrical industry's reaction to them, for the first time in Pike's history, the company began to lose money. Hit with a storm of its own, the company delved into the source of the problems. And, despite the difficulties, Floyd Pike, characteristically, looked for any opportunity.

OPEC's price increases, catastrophic to much of the rest of the world, were infusing tremendous new wealth into the oil cartel. In Saudi Arabia, King Khalid, was employing the new fortune to modernize his kingdom. In May, 1975, Floyd Pike, Joe Pike, and Haywood Merritt traveled to Saudi Arabia and to Kuwait to investigate the possibilities for work there.

With work in the company's traditional geographical service areas continuing to diminish daily, the exploration itself was beneficial. Management demonstrated, then and there, that the company would always, at least, consider, any possibility – anywhere. But, work in the Middle East was not to be; the decision itself was postponed indefinitely.

# Saudi Arabia Sets 5-Year Plan With Investment of $140-Billion

**By JUAN de ONIS**
Special to The New York Times

RIYADH, Saudi Arabia, May 18—Saudi Arabia's Council of Ministers approved today a five-year economic plan of stupendous proportions calling for a $140-billion investment.

The plan would be carried out with the assistance of a half million technicians, managers, teachers and workers, who would be imported into this country of 8.2 million, raising the labor force by 31 per cent to 2.3 million from 1.6 million. Saudi Arabia is the world's leading oil exporter.

Among the improvements envisioned in the plan are:

¶The construction of 8,500 miles of paved highways, 200,000 housing units and 2,000 miles of power transmission lines to electrify all large towns.

¶Expansion of port capacities to handle 13. million tons of cargo annually, compared with 5 million tons now.

¶Cement production of 10 million tons a year, a tenfold rise from the present one million tons.

¶Huge investments to improve water supplies throughout this desert country, including irrigation systems and easy credit for farmers.

¶$13-billion of Government-sponsored investments in heavy industry, particularly in oil refineries, petrochemical, steel and associated manufacturing industries.

¶An expansion of the educational system to increase enrollment in elementary schools to one million boys and girls from the present 600,000 and in universities to 31,200 students from 12,000.

**13.3% Growth Seen**

Altogether the plan envisages a 13.3 per cent a year growth rate for Saudi goods and services, outside the oil sector.

This means a level of investment during the coming five years nine times higher than during the first five-year plan, which was adopted before Saudi Arabia's oil revenue increased to $25-billion a year following the quadrupling of oil prices in 1973 and 1974. The purpose of the new plan is to invest domestically all the oil income that has accumulated here since the price rises.

The plan foresees a decline in the size of the farm labor force offset by a large increase in employment for construction and commerce. It does not establish the amount of oil that Saudi Arabia would produce during the plan, but the financial estimates are based on exports of 8 million barrels a day at the current price of about $10.40 a barrel.

Th plan was drawn up by the Central Planning Organization headed by Hisham Nazar, a United States-educated economist, with the assistance of experts from the Stanford University Research Institute and other international advisers.

Other foreign experts here, however, gave the plan little chance of fulfillment by its 1980 termination. Although Saudi Arabia will have enough foreign income from oil exports to finance all the projects foreseen in the blueprint for a new society, they cited social and cultural problems to be overcome.

"The new five-year plan is a good indication of intentions, but there are too many obstacles to be overcome in too short a time," said one international expert familiar with the development of the plan.

The enormous increase of foreign workers and technicians is a political and social problem in this long-secluded country, which has strongly Muslim religious beliefs embodied in its laws and social customs.

There have already been diplomatic contacts by Saudi Arabia with Egypt, Pakistan, Malaysia, and other Muslim countries to obtain skilled workers, and South Korea has offered to send large numbers of construction workers and technicians for projects here.

There is also the problem of training and motivating tradition-bound Saudi Arabian workers for the work habits and technology of a modern industrial society.

The plan document clearly recognizes these problems, but it tends to be optimistic in assessing the possibilities of solutions.

"The economic and social development of a nation requires the strenuous efforts of both the Government and the people. Despite the Kingdom's unusual financial situation, this is no less true in Saudi Arabia than in other countries, the plan document said.

Even in physical terms, the plan appears to exceed the capability of Saudi Arabia's economy to absorb the necessary equipment, supplies, and manpower.

"The problem is one of physically obtaining, moving and managing the utilization of the natural, manufactured and human resources required," said the plan document.

*My memory of Floyd Pike goes back to*
*those very beginning days.*
*In my opinion, there have been very few men like him.*
*He was very respectful of other people*
*— respectful of their feelings —*
*treated people like human beings.*
*I guess one of the very last times that I saw him*
*was down in Pine Mountain, Georgia.*
*It was very wet that morning,*
*and we were at the top of a hill,*
*pulling insulators up on a 500,000 tower.*
*I heard something coming up the hill,*
*and it was Floyd in an old green Dodge.*
*The mud was about half-knee deep, very cold.*
*The first words Floyd said were:*
*"Draughn, how's it going?"*
*"Floyd, they're the best I've ever seen.*
*I couldn't ask for it be any better!"*
*He patted me on the back and said,*
*"That's exactly what I like to hear.*
*Carry on the good work."*
*And he went on back down the mountain.*
*—Jim Draughn*

# 26

# Floyd S. Pike: Leader, Husband, Father, and Friend.

Tragedy struck the Pike family and the Pike Corporate family in 1975 at the very core when the company lost its founder. For almost thirty years, Floyd Pike's vision had inspired and led the way to achievements few would have believed possible when that first little crew set out in Old No.1, in 1945.

In describing Floyd S. Pike and characterizing his life, the Reverend Preston J. Taylor of Richmond, Virginia, a life-long friend of the Pikes' and for many years their pastor at the First Baptist Church in Mount Airy, recited:

There are hermit souls that live withdrawn
In the place of their self-content;
There are souls like stars, that dwell apart,
In a fellowless firmament;
There are pioneer souls that blaze their paths
Where highways never ran —
But let me live by the side of the road
And be a friend to man.

Let me live in a house by the side of the road,
Where the race of men go by —
The men who are good and the men who are bad,
As good and as bad as I.
I would not sit in the scorner's seat,
or hurl the cynic's ban —
Let me live in a house by the side of the road
And be a friend to man.

I see from my house by the side of the road,
By the side of the highway of life,
The men who press with the ardor of hope,
The men who are faint with the strife.
But I turn not away from their smiles or their tears,
Both parts of an infinite plan —
Let me live in a house by the side of the road
And be a friend to man.

I know there are brook-gladdened meadows ahead
And mountains of wearisome height;
That the road passes on through the long afternoon
And stretches away to the night.
But still I rejoice when the travelers rejoice,
And weep with the strangers that moan,
Nor live in my house by the side of the road
Like a man who dwells alone.[17]

Tommy Jarrell, with the Pike Electric throughout its fifty years, spoke for many, when he summarized the philosophy of Floyd S. Pike and the company he founded: "He was honest and straight as a man could be. He believed in doing everything the right way — doing a good job and being honest. He never did tell me anything but the truth."

**IN MEMORIAM**

FLOYD SWANSON PIKE
1918-1975

The death of Floyd S. Pike on June 11, 1975, was a great loss to this corporation, its stockholders and its employees. Since the beginning of the company in 1945, its growth and development have been the result of the vision and dynamic leadership of Floyd S. Pike. From a very humble beginning, he led the corporation which bears his name to its present position as one of the major companies in power line construction.

Floyd Pike will be remembered for much more than being a businessman of great ability. His employees will remember him as a friend and advisor who always had time to listen to them and to their problems. His personnel policy was well known to his associates and went something like this: "I want to treat every man working for this company as I would have liked to have been treated when I was a lineman or a groundman."

**The Directors**

# A PERSONAL TRIBUTE

*With Opal and Floyd Pike as my parents,*
*it's hard to say when I "started with" the company.*
*I literally grew up here. I have always been on the job.*
*I have never known anything else.*
*My parents provided me with a unique experience,*
*and I am very grateful. It has always seemed as if the company was another member of the family.*
*Growing up here meant that*
*I was in training with "Horse" Maynor for as long as he lived.*
*I was very fortunate to train in the field*
*under James Anderson and Ted Ashburn.*
*I got to work with Jim Watts, and for Glen Pike*
*in the Commercial Wiring division.*
*Growing up, I worked with Duke Power's Engineering Department in the summers, when Nick Shoaf was superintendent under Joe Mason.*
*When the time came for me to come out of the field into the office,*
*I knew practically everybody here — had lived with them almost everyday of my life.*
*I trained in each of the divisions.*
*When I came into the office, I already knew most of our customers.*
*They were friends of our family.*
*I feel very fortunate to have had Opal and Floyd Pike as my parents.*
*I grew up living with their dream.*
*Carrying it on was as natural as breathing.*

*—Joe Pike*

*Floyd Pike*

*Opal Pike*

*The good news is that Appalachian Power Company*
*survived those times. And well managed contractors,*
*such as Pike, also retained their core personnel*
*through this period.*
*The Floyd S. Pike Company, under the able leadership*
*of Joe Pike and Haywood Merritt, emerged from this trying*
*time even stronger than before,*
*and they seem to have never looked back.*
*By the late 1970s, Pike had become the best known,*
*most dependable, and the dominant contractor in our area.*

— *Charles A. Simmons*

# 27

# Riding Out the Storm

The most difficult storm of Pike Electric's entire history occurred in the six-month period from June 1975 to January '76. And as the winds blew hard and, seemingly, from every direction, the company lost the only leader it had ever known. In his recollections of Pike's most difficult days, Reg Banner observed that it is one thing to know that you are going to replace someone who is retiring within a certain period of time. "Then you can begin to prepare. However, when it happens so unexpectedly, it must really be tough — especially tough — when that person is your father."

Notwithstanding the personal loss, this was the only time in its fifty-year history that the Pike Company did not made a profit. "Joe Pike could not have taken over the reins as chairman and president at a tougher time," declared Reg Banner. "The times were so difficult that in fact, quite a few contractors went out of business," he said. "And many others were hurting badly."

The growing pains of the past thirty years had always been a question of how to obtain more men, more equipment, and more capital. With the incredible numbers of crew releases, personnel was reduced to around 600 employees, and the yards were filled with parked equipment. Understandably, the company was caught carrying a large overhead, and that alone was a struggle to get under control.

To add to Joe Pike's problems, when the company made calls to the Northwestern Bank to lay financial groundwork for banking assistance should the need arise, the bank's response added insult to injury. "Despite the fact of the close and long-term relationship which had existed between Floyd Pike and Edwin Duncan, Sr.," said Reg Banner, "when we contacted Northwestern Bank in '76, they sent over a young fellow who basically stated that the bank no longer considered Pike Electric to be a valuable customer." The bank was reluctant to loan Pike any money, without a high cost. Northwestern Bank had, in essence, "pulled the rug" from under the company.

"For a while there, it seemed like we got a shock a day," said Joe Pike. The impasse with Northwestern Bank was another difficulty that was totally unexpected by the Pike company. "It is a little more understandable now in retrospect," Pike poignantly observed. "I was carrying Dad's estate obligations and was new at my job; and the company did have a considerable outstanding loan with MetLife — all of this was added to the economy' s sudden impact on us." Insightfully, he concluded, "And, after the death of Edwin Duncan, the Northwestern Bank itself was in the process of learning to live in a new era."

It is a matter of record and a point of historical significance that at no time was the Pike company ever at a point of reorganization or receivership. "It was just the first time in our history that external problems had halted our growth and progress," Joe Pike said, "and growth and progress and Dad's leadership were all that the company had ever known."

Another disappointing but understandable loss occurred at this time. Robert Schwartz resigned from Pike's board of directors when he was appointed Met Life's chairman, CEO, and president. Fortunately, however, the Pike loan-account with Metropolitan became the responsibility of Robert Chappell.

Chappell, senior executive-vice-president and chief investment officer of MetLife, now retired, was a junior investment officer on a family vacation in Georgia with his wife, Linda, and their children, at the time Met directed him to call in Mount Airy. Chappell was already familiar with Pike, having visited the company in the early '70s, during the time of its meteoric growth. His specific instructions were to "spend as much time as needed for evaluating Pike's situation, to recommend necessary steps to return the company to prosperity, and to protect Met's investment, which by that time had grown to a second loan." His new assignment brought him back to Mount Airy where he found a team hard at work, putting the pieces together in a new way.

"From the first meeting," Chappell said, "the quiet confidence that Joe Pike exhibited was surprising. Here was a person thrust into a position of leadership years before he expected it, having just lost his father, with a company suffering the impact of a weak economic environment, and seemingly nothing going in the right direction." In Chappell's opinion, although Joe Pike was new at running the company, he confronted the problems head-on, and by listening and respecting the opinions of others, created a team approach that was correct for the times. "Joe Pike's vision of what the future could be," said Robert Chappell, "complemented perfectly, the practical experience of the others." Chappell added, that despite the grief and the personal obligations that Joe Pike faced, "even then, he was unrelenting in maintaining the company's dedication to its customers and its reputation for service."

Chappell visited the Pike company often during those troubled times and shared his advice and his support. He also, very professionally, disguised any doubts that he and others at MetLife had about Pike's recovery. "In looking at Pike today, I am amazed that, in the mid '70s, there were questions about the company's future." He added that, in retrospect, MetLife underestimated the company's potential to emerge from its problems so successfully.

"For many years now," Joe Pike commented, "we have had only limited transactions with MetLife, but Bob Chappell has continued to generously share his advice and the benefits of his experience." Pike said that, from the start, Chappell always seemed to be cheering for the Pike company as the "home team." Indeed, Robert Chappell remarked that working through this difficult period with Pike was one of the most satisfying memories of his career. "I don't think that we could ever ask for a more loyal friend, personally and corporately," said Joe Pike.

In the toughest days in '76, Joe Pike's leadership was tempered with the patience that the disrupted environment demanded, but, characteristically, his eye was on the future as he assessed the company's situation. "Many of these cutbacks, by their very nature, should be viewed as delays of work that must eventually be done, if reliable service is to be maintained," Pike wrote to the shareholders.

He had been well-taught by his predecessor to look for every opportunity. "The current slowdown can be viewed as building a backlog of work that must be done sometime in the not-too-distant future." Considered from a calmer perspective, the company was still quite financially strong; although it had experienced a year that, in Joe Pike's words, was "not the rapid, advancing type we were accustomed to."

On the plus side, the Commercial Division had completed a very profitable year. "And to the company's advantage, Pike Electric is well-equipped to accommodate a considerable increase in work without the corresponding huge costs of machinery and equipment that is usually associated with such an increase," Pike pointed out.

Despite the slowdown, Joe Pike reminded that "the

company still had its most important asset — the integrity for which it is known. "Moreover, our company, owned by its employees, is in the enviable position of stockholders and employees, being one and the same, and having identical goals." So the company dug in to ride out the storm.

It was imperative to make every effort to keep as many of Pike's personnel as possible. Management took the initiative to avoid any more layoffs than were absolutely necessary. All of Pike's salaried personnel unanimously agreed to a pay cut, which increased in size, up the organizational chart. And, despite the ostensibly overwhelming difficulties, the company was functioning as a team.

"I think that Joe Pike brought our team together," Reg Banner said. "We began to communicate more with each other, with the middle managers, and the supervisors." Under Joe Pike's leadership, regular Monday morning meetings were scheduled which improved communication between departments and divisions. "We then knew what was going on elsewhere in the company," Banner explained. And quarterly meetings were established, so that the company's supervisors, who are almost always on the road, could be present at a regular session for dealing with matters of policy and safety and problem solving, keeping them better informed and closer to management. "Joe Pike is probably the fairest person I've ever been around," Banner said. "He is very much aware of not treating someone or some group of employees differently from another. All of these measures made the members of management feel like we were indeed a team — that we all counted."

"From this base of unity," predicted Joe Pike, "and with the continuing teamwork it has always had, our company is ready to advance and to obtain, at least its share, of the future growth that is coming to the electrical utilities."

Even in the face of disappointment, Pike's confidence and determination produced a spark of optimism. The company marked another point in its maturity. It was going to come out of its difficulties with its team and its most important attributes firmly in tact. Pike's eye was focused on its future, and its vision was good!

# Climbing Up...and...Up ...and...Up...

*Certainly Floyd S. Pike is one of the leading*
*companies of its type in the world,*
*and First Citizens is proud to have been privileged*
*to serve as your bank for three decades.*
*If my memory serves me correctly,*
*it was in the '70s when the management*
*of Pike Electrical Contractor, Inc.,*
*and some of our senior officers met in Raleigh*
*to discuss a banking relationship.*
*Based on those early meetings,*
*we realized that Pike was a well-managed*
*company with a vision and an understanding of its business.*
*Fortunately, we had the wisdom*
*to recognize this as an outstanding*
*business opportunity for us,*
*and we were able to respond to your banking needs.*
*It was one of the best business decisions we have made.*

*—Lewis R. Holding*
*Chairman, First Citizens Bank*

# 28

# New Ways

A bank and its services are essential to a company's operations, and Pike had to find a new one. To obtain the corporate financing necessary to get the company "over the hump," Joe Pike and Emmett Forrest approached the First Citizens Bank. Richard E. Maxey was then the bank's regional investment executive, and he was well aware of the utility industry's reverberations and fully informed of the Pike company's immediate difficulties. Nevertheless, having been taught, he said, that "many times there is more in the man, than there is in the land," Maxey decided to assist Pike in seeking the corporate banking relationship that the company needed.

Without the "greatest hope of being able to fund this credit," Richard Maxey submitted a loan package to George Broderick, president of First Citizens Bank at the time.

After a wait of 91 days, Richard Maxey was informed that the bank would fund a loan. However, he had been told by George Broderick that "the figures certainly did not support the request!" As a matter of fact, the bank's credit analyst had actually decided at first that the loan would not be a good credit. "The management of the company, along with the business opportunities that lay ahead," Maxey explained, "had convinced the analyst that Floyd S. Pike would have the ability to repay this credit and become a good customer of First Citizens Bank. Without a doubt, the character of the management turned the tide in the bank's decision."[18] George Broderick concurred. He said that First Citizens Bank's decision was indeed based on "Pike's leadership" and the fact that " integrity is a hallmark of Pike Electric."

The loan and the confidence that the bank exhibited in Pike seemed almost to jump-start the company's recovery. "I dropped by and carefully monitored the direction of the company," said Richard Maxey, "and amazingly to me, it seemed like it almost picked up speed, just a little bit, but picked up speed from day one!"

"Strangely enough," Reg Banner said, "in those early years with the bank, we borrowed relatively little, from '77 through '80, and we did not need to borrow from them again for almost fifteen years." And during those years, the decision that George Broderick, Richard Maxey, and, ultimately, Lewis R. "Snow" Holding, chairman of First Citizens Bank, had made to provide corporate financing, allowed the Pike company to play a major part in the growth of First Citizens Bank in Mount Airy.

The local bank was the "new bank" in town at the time, and Cooper Adams, First Citizens executive officer in Mount Airy, had just become its new manager. "The association with Pike gave us immediate legitimacy in the business community and an opportunity to develop personal banking relationships with members of the Pike management team and their line associates." Consequently, Cooper Adams said that this was "the one association which would ultimately insure the bank's success in Mount Airy, and it was a situation that allowed both businesses to prosper." Today, the local First Citizens Bank serves hundreds of Pike employees.

Chairman Lewis "Snow" Holding said, "I know of no relationship that our bank values more than that of Floyd S. Pike Electrical Contractor. We have worked together, and together, our companies have grown in size and strength. It has truly been a win-win situation." Chairman Holding added that, in '91, First Citizens Bank requested that Joe Pike serve on the board of directors of "both the Bank and its holding company," citing Pike's "talents as an executive" and the Bank's respect for him as an "individual of great character. "This is a rare combination," Holding commented.

Summarizing the working relationship that continues now after almost thirty years, George Broderick said, "Let me comment on the relationships with the company and with Buddy, (Joe Pike). Some of this is corporate and some is personal!"

"Annually, Joe Pike invited — and sort of insisted — that the top officers of the bank meet with top officers of Pike Electric to review the results of Pike's fiscal year and its plans for the current and future years. In this way," Broderick explained, "the bankers felt that they were an integral part of the company." He added that, in his opinion, "This was a most revealing and important meeting and something that every large company should do."

George Broderick said that after only a few of those meetings, he began to realize that Joe Pike had another most important motive. "It was an opportunity," Broderick said, "for the bankers to meet and get to know other members of his management team and to assure that there was a continuity of leadership if something unforeseen were to happen to him."

On a personal note, Broderick added that after he retired, he received an invitation from Pike to attend one of the review sessions. "This was one of the nicest compliments and most pleasant experiences, in my affiliation with Buddy Pike as banker, fellow director, and friend."

*One of the most beneficial aspects of Reg Banner's work*
*with this company came at a time when we really needed it.*
*Through the accounting procedures that he had in place,*
*the company could know both the cost*
*and profitability of every single job.*
*In the euphoria of the early '70s,*
*when transmission work was strong,*
*we were so busy doing the jobs,*
*that the cost of doing them*
*had not been monitored closely enough.*
*But with the slow down — we were*
*attempting to plug every drain!*
*Reg's numbers gave us the information to decide*
*to discontinue large-scale transmission work.*

*—Joe Pike*

# 29

# Plug Every Drain

During the slowdown of the 1970s, Pike took a hard look at itself to determine its future. When the numbers were analyzed, a major decision was made to end large-scale transmission work. The decision in itself was indicative of new ways of thinking and new ways of operating — different from those of the past, but certainly appropriate for the economic environment that was emerging. Pike's "can do" attitude was in no way intimated. However, from then on, anything that Pike "could do " would be done, only after wise and careful consideration. "This was a monumental business decision made by Joe Pike and the management team," Reg Banner declared. "This was company's business philosophy of the future."

As transmission work was scaled down, the company was still saddled with the EHV equipment. In April 1977, Pike contacted the Forke Brothers in Nebraska to conduct the largest auction the company had ever had. A vast display area on a mountain-top surrounding the Ararat Rock Products quarry was provided by its owner, J.H. Crossingham, Jr., president and chairman of Spencers's Inc., in Mount Airy. Over 300 pieces of huge transmission equipment were readied by the garage personnel, who worked tirelessly and set a standard for themselves in auction preparation that has been maintained and refined in every subsequent auction.

**THRONG OF BUYERS** and onlookers crowd around auctioneer at sale of construction equipment by Floyd S. Pike Electrical Contractor Contractor, Inc.

*Forke Auction*

By the day of the sale, air traffic at the Mount Airy-Surry County Airport was at an all-time high, and many of the wives and family members of Pike's personnel volunteered to serve as guides and to provide shuttle service from the airport to the auction site. Buyers were registered from 22 states and Canada.

Logistically, the sale was a successful team endeavor, and the profits were visible — in morale and in dollars — both of which provided long-term benefits to the company. Following the sale, Joe Pike and Haywood Merritt negotiated an agreement with MetLife so that the Pike company could apply the net proceeds from the sale to the MetLife loan as an extra, early payment. "I think that this measure really speaks well for Pike's management," said Reg Banner. "Although the company was struggling to make a profit, we elected to reduce our debt and our interest expenses, so that we could begin to get our financial store in order for the upcoming years of growth and prosperity."

## TARGETING ADDITIONAL SAVINGS

For additional savings, in terms of both lives and dollars, Pike's management team targeted its own safety program. Glen Pehl, president and CEO of McNeary Insurance Consulting Services, Inc., said that when he called on the company in the late '70s, he found a management team "not just concerned about the monetary effect, but they were more concerned about the excessive pain and suffering of employees which could be brought on by an ineffective safety program." Safety, in general, had been thrust to the forefront of importance when the company lost two of its employees in a plane crash in 1978.

John Thompson, Pike Electric vice-president, and his assistant, Jerry Chinault, crashed on approach at the Homestead, Virginia airport. With a love of flying and piloting his own plane, Thompson was next in line to land, following an FAA plane carrying the agents who were to make a final inspection of the work Pike had just completed there: re-lighting the runway.

"About 80 percent of Pike's airport work was upgrading existing facilities," Claude Brinkley said. " We had several small airports — like the one at Homestead — and we had the runway project at Roanoke." In addition, the company had some very large airport projects, including the re-lighting of a major part of the Charlotte Airport.

"When John died," Brinkley recalled, "I wound up with the task of trying to determine whether we should stay in the airport lighting business or forget about that type of work." He said that when he was questioned by Emmett Forrest at a regular Monday morning meeting about the safety and liability involved in those jobs, management decided that the work did not justify the risk.

Despite all that Pike's management team had recently faced, Glen Pehl said that when he came to discuss insurance matters in '79, he found a well-focused committee that "kept him on his toes with their astute questions and observations." This quality impressed him and dispelled any worries that he and others in his company had about Pike's future.

Nonetheless, Pehl termed Pike's corporate safety program at that time, only "passable." Of immediate concern to the company was the fact that it was facing a stiff worker's compensation experience-penalty. "Based on the payroll for that year, this meant that your company would pay out an additional premium which was in excess of the average for your particular industry," explained Glen Pehl. Given the facts, Pike's personnel studied the situation carefully to determine the appropriate remedy. "I have always appreciated the attention, thoroughness, and Pike's constructive reactions," said Glen Pehl, commenting on Pike's approach to any problem.

Constructive reactions were forth coming, "although during the early '70's, Pike was already ahead of its competition in promoting safety," said Zack Blackmon, Jr., "We already had mandatory requirements of wearing hard hats and rubber gloves and sleeves when working near energized lines." Pike's requirement of wearing rubber overshoes while working around energized power lines is still a unique safety requirement of the company.

Nevertheless, when the numbers indicated that further initiatives were mandatory, A. L. McAlexander and his Safety Department really went to work. First of all, he developed a disabling-injury index that identified problems and indicated areas for improvements. He strengthened his department further by appointing Robert Chilton safety supervisor. "I remember telling Robert Chilton," said A.L. McAlexander, "that with help of the good Lord, and a lot of people, we were going to do something to reduce the serious accident rate in our company!"

Right away, Robert Chilton went on the road teaching first-aid to Pike's linemen. Along with his other responsibilities, he was also in charge of the company's dielectric testing, making the necessary exchanges in rubber hoses and blankets in the field. Mark Sams was placed in charge of the department's testing vehicle. Later, Bobby Gammons took over that responsibility.

*Safety Supervisor conducts a safety class for garage personnel. L-R- Fred Smith, Larry Good, Roger Jones, Charles Ward, Doug Hooker, Luke Easter, Jimmy Romer, James King, Ronnie Collins, Joey Mitchell, James Tillotson, Robert Chilton*

The efforts were company-wide, and demonstrable results were quickly evident. "The change was brought about mostly when Joe Pike began the quarterly-scheduled officers/ supervisors meetings," said A.L. McAlexander. "Then and there, safety became an integral part of the agenda." Consequently, the message of on-the-job safety was carried back to the field, to the front-line supervision, to the foremen and to the individual employees. "Our disabling-injury index fell to a fraction of what it was at the beginning of our directed effort," McAlexander said.

*Load testing at main garage, Mount Airy, North Carolina*

And the savings were evident in many ways. Glen Pehl — who has since called the company's safety program, "now second to none" — summarized: "This effort has made Pike a much safer place to work and has saved your company hundreds of thousands of dollars in insurance premium reductions." Another very important result of the company's emphasis on safety is that it "has allowed your company to obtain more work through the competitive-bidding process, due to lower insurance premiums," Pehl pointed out.

Indeed, the efficiency that has resulted from Pike's safety efforts has not gone unnoticed by its customers. "As far as I can recall, in the 35 years that Pike has done the work here for us at Satilla EMC," said Leman Lanier, "there hasn't been a single accident of any serious consequence." What makes that fact even more remarkable is that the Satilla system is about 95 percent complete in its conversion from 7200 to 14.4. "That is some of the most hazardous work you can do in the electrical industry," Lanier explained. "And to do as much as Pike has done, with that accident record, it is just almost unheard-of!"

## TARGETING SAVINGS WITH TECHNOLOGY

The company looked to technology to implement additional savings. In 1975, the company bought its first mini-computer. Joe Pike and Reg Banner went to Winston-Salem to see a demonstration of an NCR-399 and bought two of them. When they arrived, Pike began developing a job-cost system. "Before computers, there was no cost accounting as we know it," said Richard Isaacs, Cost Accounting Manager. "What passed for accounting, was done on an 8-1/2x 11" ledger sheet and covered only the major jobs."

Isaacs, a field accountant in Louisville and Cincinnati for a large chemical-plant contractor before he came to work with the company in 1971 — only a few months before Reg Banner — was at first in charge of the bookkeeping for the Mount Airy-Surry County Airport, until the company relinquished that responsibility in '75, and for Pike Equipment & Supply Company. He and Banner established an immediate comaraderie, and they discussed accounting plans and systems for the company's future for hours at a time.

It was almost a year after getting the computers, before everything was in working order to produce the first job-cost report. Even then, the company's first system was really crude when compared to today's, and most details had to be computed for each job manually. "The system required three people who worked furiously," said Isaacs. "Reg Banner, Johnny Mullins, and I had to sort each job's revenue, labor, equipment, material, and overhead, and match it to the financial statement each month." A task that sometimes caused Isaacs to work all night long.

Banner described Isaacs as an ideal cost accounting manager. "He is so easy going and does not get upset when a manager comes in questioning the profit on his jobs at the end of the month," said Banner. "And Richard's contribution just keeps on giving — his son, Keith, is personal computer manager in our Information Systems Department. His other son, Kevin, is an underground foreman in Pike's Northern Division. We owe much of the success of our Cost Accounting Department to Richard Isaacs."

When Pike bought its first true computer in 1978, an IBM system 32, the company began to automate and computerize the garage operation in accordance with the Vehicle Maintenance Reporting System (VMRS), which provided an up-to-date garage inventory and a ready indication of equipment cost. "The software — the inventory and equipment-cost program — was set up according to the American Trucking Association's Vehicle Maintenance Reporting Standards," explained Joe Reid, Director of Information Systems, "and these standards are still currently used."

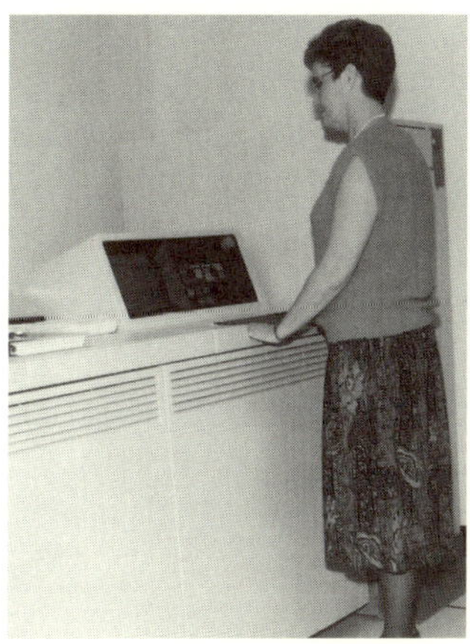

*Shelby Taylor*

With a background in industrial engineering and data processing, Reid carried out the task of setting up the newly-created computer department, and he worked as liaison between the garage and the accounting department to coordinate a smooth installation.

The company's IBM system/34 arrived in October '78, and went "live" in March '79. The first data entry operators were Libby Nunn and Shelby Taylor whose responsibilities were to set up and maintain all parts and equipment files.

Computers expedited Pike's general accounting responsibilities and the accounting associated with the company's retirement plans for John Mullins, Pike assistant secretary and assistant treasurer, who was hired from Ernst & Ernst in '73, while he was working on the Pike audit. "With Johnny, we have an in-house expert on financial statements, profit sharing and 401(K) plans," said Reg Banner.

On a personal note, Reg Banner said that in the days before computers, Floyd Pike was forever requesting financial data that was "very time consuming" and "difficult to put together." On those occasions, Banner, John Mullins, and Richard Isaacs would then "set up shop" in a vacant office and work as long as the task required. "When we opened those doors and came out of there, the three of us usually had the information surrounded."

The establishment of the Information Systems Department is considered by many to be one of the pivotal points in the company's history. "Without this department," observed Richard Isaacs, "it would be impossible to furnish the information that is needed to be as competitive as we are, or to handle the volume of business with as much detail that we do today." Isaacs commented that Pike's competitors certainly have data processing, "but I will venture to say, that they have not applied theirs as effectively as we have ours."

With up-to-date technology supporting effective accounting, the end of the decade was near, and the company's financial store was in order — anticipating the future that was quickly coming. Pike's share of the rebound in the utility industry was, indeed, on its way!

*Accounting was kept in the ledgers when John Mullins and Richard Isaacs came to Pike in the '70s.*

*"The three of us usually had the information surrounded."*
*— Reg Brown*

*In all honesty, I began work with the expectation*
*of saving a few hundred bucks*
*and doing some more traveling around the country.*
*I didn't count on line work being like the circus.*
*It has a tendency to get into your blood.*
*It is a good feeling to see a pile of hardware turn*
*into something you can be proud to say that you did.*
*I didn't count on meeting people like Mike Blackmon, our*
*foreman; Reid Clifton, the first man I ever crawled into a yellow*
*truck with;*
*Robert Thomas; Gene Cook; Lyonel Atkins;*
*Pop Lipscomb; James Paul; and "Shorty" Jessup;*
*and many others from whom I learned line work—*
*mostly at night, when power lines were built on a grand scale*
*in rented rooms and bars. The hard people, on many occasions,*
*weren't quite so hard. As you can probably guess,*
*I feel that it is the people of this company that make it what it is.*
*These people furnish us with the best equipment*
*and the latest tools in order to best perform the job we do.*
*When you have a great balance between the people who get the*
*work and the people who do the work,*
*success is almost guaranteed.*
*My hat is off to all the people*
*who have worked hard to make this possible.*
*I am especially thankful for the leadership of*
*Opal Pike, Buddy Pike, Haywood Merritt,*
*Zack Blackmon, and the many others*
*after the untimely death of Floyd Pike.*
*Without their dedication, under trying times,*
*it could have been an entirely different story*
*—Charles M."Droopy"King*

# 30

# Poised and Ready

During the later half of the decade, the company was, indeed, emerging from the slowdown. Difficulties had been endured. Lessons had been learned. Appropriate actions had been taken. During its period of introspection, Pike's management team had examined virtually every aspect of the company. Pike's storm procedures were no exception, and the company stayed prepared. This was a good thing, for nature was unaware of an industry slowdown.

The winter of 1978 was brutal to Illinois. When Pike received calls for help during an ice storm, Claude Taylor, Frank Jessup, Tommy Jarrell, and Milton Bumgardner went to oversee the restoration. Field mechanics Ted Atkins, Clinton Hal, and William Chappell were on the road, as well as were Jack Arrington, Carl Bailes, Frank Barger, Latney Childress, and Joseph Hall, who went to assist the Menard Electric Co-op at Petersburg, Illinois. Larry Hall, Ronnie Martin, James McPeak, John Nipper, and Milas Palmer had their crews working in the same area, together with the crews of Earnest Pollard, William Rasnick, Gary Robinson, Audie Simmons, Steve Thornsberry and Ronald Tucker. Ed Brown, Jack Carper, and Troy Snead took their crews to the Corn Belt Co-op at Bloomington, Illinois. And William Coffey, Mitchell Hunter, Sammy Vaden, and J.W. Wright went to Auburn, Illinois, to assist the Rural Electric Convenience Electrical Cooperative.

All in all, Pike was moving along quite well — but carefully — in a positive and disciplined procedure. All expansion and on-going projects were scrutinized from every angle in order to provide quality in service characteristic of the company's reputation. Claude Brinkley explained that this in no way hampered the company's momentum. "The expansion we had in the early days, and to some extent today, has a lot to do with our reputation that has been created by our ability to get the people — the right people — the equipment, and the supervision together so that we start and complete projects on time." Brinkley said that the new policy demonstrated that "we did not want to expand into any areas that jeopardized or interfered with our existing obligations — which were to supply our customers with the best of everything we had, with 100 percent of our efforts."

This policy presented a brief dilemma when, for the first time, Columbus and Southern Power Company, Columbus, Ohio, called in the late '70s, in anticipation of an impending strike on its system. In responding, Claude Brinkley explained that Pike would provide help during a strike, only if the company were already on the job. Mother Nature then played an ace for both parties involved. An ice storm occurred, and Pike was called for help. Consequently, the company was not only on the job during the storm which hit as the strike occurred, but the assistance resulted in work that has been continual through the years.

Recovery was coming in familiar form. Union activity was very strong during the late '70s, especially in Kentucky, West Virginia, and Ohio. Audie Simmons, assistant vice-president, Central Division, said that he carried his crew into Huntington, West Virginia, in 1979, when AEP had power outages on their system where poles had actually been cut down with chain saws. "My crew left Ashland, Kentucky one evening, and we were met at the West Virginia state line by an AEP official who carried us to the area where the damage had been done," Audie Simmons said. "We worked the restoration all night, and we had planned to be out of there by daybreak to avoid the strikers." However, Simmons explained that "the work area was in a hollow between two mountains, and as day break came, and we climbed out to the main road, it was lined with strikers."

He said that tension mounted as an American Electric Power official escorted them safely to a truck stop for breakfast. "But while the crew was eating," Audie Simmons said, "the parking lot there began filling up with more than fifty union workers." He said that his ten crew members, along with Richard Simmons and Claude Taylor, were greatly outnumbered. "This situation was about to get nasty!" Audie Simmons said. "The strikers started vandalizing our equipment and making threats to us." He explained that the training and discipline of Pike's supervisory personnel were unforgettable. He said that Richard Simmons and Claude Taylor remained calm and called a sheriff, who safely escorted the men into Kentucky. "I surely was glad that Dad and Claude Taylor were there." Audie Simmons said. " I was very impressed by the way they handled the situation." He added that this situation only prepared him for difficulties that came the next year when his crew was building a transmission line at the Big Sandy Power Plant in Louisa, Kentucky.

"Although the Kentucky union workers were not as aggressive as the West Virginia strikers, they still let us know that they were there when they gathered at the power plant each day," said Simmons. At the entrance gate, the strikers had set up a big billboard, and whenever a non-union worker crossed the picket line, his name was soon featured. "One morning as we passed the entrance gate, the billboard read, 'Audie Simmons and Pike Electric — Scabs of the Day.' " Simmons said that was not the way he had envisioned someday seeing his name in public. "But it was quite a learning experience."

Utility officials were also impressed with Pikes' courage and calm methods. A letter received from John Vaughan, then president of Appalachian Power Company, confirmed that the efforts made by the men of Pike made during these tense times did not go unnoticed. "I, particularly, remember in the late '70s, when the power company was dealing with strikes at a number of locations. Consistent with Pike's willingness to always help, your crews were sent into the Beckley area to help us meet our public-service obligation." He explained that this was another example of a "volatile situation in a tough labor climate, without any police protection.... It was an example of your willingness to assist us during the most difficult of times."[19]

*Jobs often require "cool heads" in tense situations.*

*"Akers, I want five underground crews*
*in Hendersonville on Monday." —Zack Blackmon, Sr.*
*"Zack! Do you realized that today is Thursday?"— Akers Woods*
*"Hell, yes, Akers! And tomorrow's Friday !"*

# 31

# The Future Is Finally Here: The 1980s

As the 1980s began, Pike was growing, but growing very carefully. Company minutes record:

> The Board of Floyd S. Pike, Electrical Contractor, Inc., met at the corporation's offices in Mount Airy, North Carolina, at 11:00 a.m., August 25, 1980. A discussion was led by Joe Pike concerning the long-range plans for the company. The general conclusion was that, although now is not the time to do it — but in the near future — the company is going to need to make some considerable investments in additional equipment, a new garage building, or an addition to the existing garage.

It had been a very good year. "It only took us four years to really turn things around but good!" said Reg Banner. "In 1979, we saw our profits soar by more than twice the profits in our 'great and historical' year, 1974." He added that in "only five years after Floyd Pike's death, I was able to re-live one of my favorite memories — that proud moment back in '74 when the company had finally made a million dollars!" Reg Banner said that the flashback occurred standing in Joe Pike's office one day in 1980, "and Buddy discussed the fact that Pike had reached another milestone: the company's net worth was over ten times that of 1974."

The *Pike Line* reported: "Your company has had a very successful year ending, June 30, 1980, and your management is very optimistic that the coming year will also be a good year."

Although in the summer of '81, the economy, in general, experienced a slight recession that slowed the company's work a bit in some areas, Pike's growth continued through acquisitions. On July 4, 1981, Pike purchased the Georgia Right-of-Way Contractors, Inc. and renamed it, Georgia Right-of-Way Company. Jennis Everette was the area supervisor at the time of purchase, and the new company was already under contract to provide tree trimming and bush-hogging in right-of-way clearance to Pike's customers, Georgia Power Company, North Georgia EMC and the Coweta-Fayette EMC. Before the decade ended, Georgia Power Company recognized the Georgia-Right-of-Way Company for excellence in its work.

In 1982, Pike's presence was strengthened in Kentucky with the acquisition of the Harris Brothers Construction Company, located in Prestonsburg. Harris Brothers' work was mainly reconductoring and conversion in a nine-county area — operations and locations that fitted well into Pike's existing service area for Kentucky Power Company. The new operation was under the supervision of Pike's general foreman, Phillip Anderson, who was in charge of the Pike crews serving Kentucky Power's Williamsburg-Sprigg Division. Bobby Inman was general foreman over crews at the Pikeville Division, and Posey Edwards, who had worked with Claude Taylor since the Pike company's first job in Kentucky began, supervised his crews from the Hazard Division.

## THE WORK INCREASES

By 1982, Pike's personnel numbers had climbed to an impressive high of 1,600, and equipment numbered 1,400 pieces. Jobs increased in numbers and in diversity.

During the '80s, Pike's Central Division grew from 200 to 1,000 employees in the Carolinas alone. Barny Ratliff discussed the strategy behind this growth. "At this time, I joined supervisors, Ralph Anderson and Jack Farris, and we began to call on potential customers. Ratliff said that Anderson was very good at introducing himself — and the company — but Jack was more reluctant. "However, if someone else opened the door for Jack, he would walk right through it." The results of the efforts of these men became evident when Anderson's responsibilites increased to the point that he needed assistance which was provided by Paul Greenhill who was appointed area supervisor during this busy time.

Working under these men, Pike's general foremen directed crews that provided the full range of the company's capabilities. Romey Clement's crew was constructing new sub-stations and rebuilding others. Weldon Jessup and Jack Arnder were assisting Lewis Goins's transmission crew with energized relocations, and Junior Goins and his crew were building new 100 kV transmission lines. Crews were doing hot-stick construction of 69kV, new overhead, underground, and reconductoring. Bill Koone, then a new general foreman in the Central Division, had his crews were working out of Shelby, North Carolina, and Thurman Rector moved into Hickory, North Carolina

Pike crews were involved in a great deal of work that was being done in Charlotte where the city was in the throes of a major downtown beautification effort. All of the city's sidewalks were widened which required the relocation of every street light on the project. And Pike's underground crews were assisting Duke Power wherever duct banks, manholes, and vaults required modifications.

Theodore Helm, then an underground (URD) foreman with seventeen years of experience and a reputation for landing the toughest jobs that the company had in Charlotte, was assigned the task of re-routing Duke Power's facilities. His stamina, as well as his skills, were tested again and again by the tedious work. Beneath the city streets, Duke Power shares space with telephone, water, sewer, and gas companies so that the relocated electrical facilities had to coordinate with that of the other utilities, while dodging the existing runs and associated service laterals. Also, at the same times and places, the city's streets were being resurfaced, which added more complications — with paving contractors working against the clock to meet their deadlines.

In the Carolinas, some transmission work was being done. James Anderson and his crew were building the Seneca to Oakway 44kV for Blue Ridge Electric at Pickens, South Carolina. Jasper Marion directed Donald Hiatt, Dean Browder, Bill Brannock, Bill Wagoner, Mike McHone, Bill Ramey, Jeff Puckett, Romey Clement, Pete Welborn, and Charles King in Washington, North Carolina, where they constructed a substation that was one of the largest in the southeast.

Pike went to work again for Carolina Power and Light in Raleigh. Audie Simmons led the build up of the operation, which quickly became one of the company's fastest-growing territories, requiring a satellite office and garage. Kenny Taylor continues to run the company's operation, and today, the facilities also include a warehouse where an inventory of material necessary for CP&L's work is maintained.

In Georgia, Robert Buffer took his crew to Newnan, where they began additional work for the Coweta-Fayette Electrial Membership Co-op. Crews continued to serve Georgia Power company.

Ronnie Crowder's crews began transmission construction for the Lee County Cooperative in Fort Myers, Florida, and his crews also had new work, that included cable, for United Telephone at Naples, Florida. The upturn in Florida led to

*Pike facility in Raleigh, North Carolina*

new jobs in the Eastern Division of Florida Power and Light Company so that Alfred Higginbotham moved his crews into South Bay and Belle Glade, Florida.

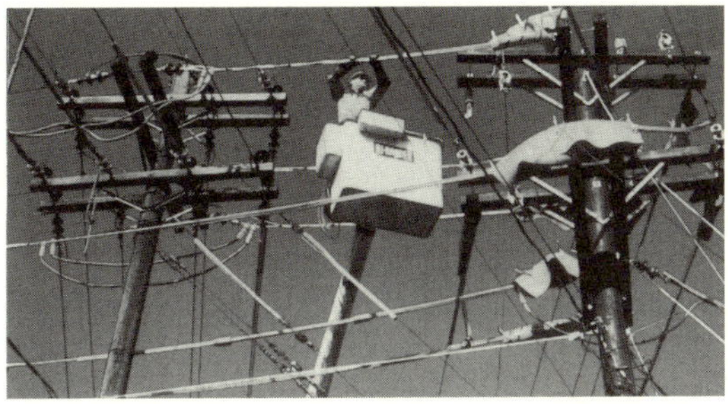

*Crews often work among many energized power lines.*

At Deland, Florida, Pike Foreman Chip Callahan had his crew on a pole-replacement job that included two-and-a-half miles of reconductoring in one of the city's high-traffic areas near Stetson University. The project became really interesting as the crew was working on, and around, as many as eighteen energized conductors in one location.

In 1986, Pike crews were summoned for unusual duties in Tennessee. Jerry Wright's crew, while working for the Athens Utility Board, painted the columns at the McMinn County Courthouse, hung huge welcoming banners, and connected the power to be used for television during the visit of President Ronald Reagan. Wright, and the other crew members, Gerald Wilson, Ronnie Hillard, Danny Bingham, and Wendell Ward, were given front-row seats during the President's speaking engagement. Soon after the visit, the company received a letter from President Reagan acknowledging the assistance that the crew had provided and thanking them for their efforts.

Besides the temporary work in Athens, new contracts were obtained in Tennessee with the Cumberland Electric Co-op at Clarksville. And in Virginia, Pike returned to work for VEPCO, while crews continued jobs for the City of Martinsville, Virginia

These were also exciting years in the Commercial Wiring Division. Doug Perkins, Jarrell Francis, Doug Pauley, and Tom Senter were supervising increasing numbers of wiring projects that included schools, libraries, supermarkets, laboratories, fast food franchises, complex wiring in industrial plants, and assisting line crews when needed.

And with the increase in work, a new building — called Pike North — was acquired for the Commercial Division. Today, the wiring and the HVAC share the larger and more convenient space which is located on Riverside Drive, approximately one mile from the corporate headquarters in Mount Airy.

Butch Riley recalled that at the time of the move, the division was encouraged to "get modern" when Joe Pike brought a personal computer over, and placed it on his desk and suggested that the department "start using newer technology!" "We suddenly had more room," Butch Riley said, "and Joe Pike suggested that we purchase a computer-controlled duct-layout machine to go in it." Riley said that he thought the new machinery would reduce cost, and it did; but an unexpected benefit was the tremendous increase in production. "The shop could now supply five times the field crews that we used to, really enhancing our possible total revenue."

*Pike North.*

## DEMANDS ON THE SUPPORT SERVICES

Facilitating the expanding work of the Pike company in the early 1980s was the store of equipment that the company had kept standing-by awaiting new line work whenever it occurred. The garage personnel had repaired and readied the equipment for the upturn. During the slowdown of the '70s, and during any slow periods since, the garage has used its time productively. "The slowdowns and recessions that have come our way, since I've been here," said John Brewington, vice-president, Fleet Operations, "have been times to prepare for the next upturn — evaluate equipment, sell what needs to be sold, rebuild those pieces needing rebuilding, and make repairs where needed." He said that when he came to Pike in '81, he found a management team "poised for expansion and a situation where managers were doing their jobs with minimal interference. The management was very focused on the needs of our customers, and there was a strong sense of family in what everyone did."

The company's garage personnel has a long-established reputation for superiority in the electrical industry, and during the '80s, the garage operation gained additional recognition when it began its own remounting and rebuilding service. Today, Pike's garage has on-site capabilities to totally rebuild derricks, transfer them to new chassis or truck bodies, do customizations, and re-install hydraulics. "We try to determine what is the best equipment for each job," said Akers Woods. "If the equipment is not available or is not on the market, then we will build it in our garage." For example, Pike's four-wheel-drive pickup — designated a 38H— is a modification of the old power wagon. "And we modify our 37W's to be used on underground," Akers Woods said. "This modification service is provided to Pike's customers, if a particular need can be met."

John Brewington said that all of these initiatives that were begun in the '80s, including the standardization of Pike's fleet, "were done with an "eye toward cost and value-received. In my opinion," Brewington observed, "good value was received for the work that we did during these times." And it continues.

As was predicted, the increasing demands placed on the garage personnel and their work space required the company to build a new 26,000 square foot garage on the main-office campus in '84. And in Prestonsburg, Mervin Blanton moved his personnel into a new 6,000 square-foot-garage completed in the fall of the same year.

In Monroe, Georgia, the Pike operation there was outgrowing all of its facilities. A new office was completed in '84, and additional garage space was required. By 1985, there were 320 line department employees in the Southern Division, and within two years, the work force had grown to 657 men on 115 crews. Six supervisors were then required, and James Dubose was named a full-time safety supervisor there. In North Carolina, garage and office facilities were added in Asheville and Wilmington; and in Florida, at Fort Myers.

*Fort Myers, Florida.*

To relieve some of the workload on Jim Draughn and his Supply Department's headquarters in Mount Airy, warehouse space for a Supply Department was added to the new garage in Monroe. And none to soon! Today, the warehouse there distributes supplies for as many people as the main warehouse in Mount Airy did twenty years ago.

Demand on the support services requires keeping adequate and anticipated supplies, tooling-up the trucks, and getting anything else that a crew or a satellite office needs shipped to its destination. This is a logistical feat accomplished by a Supply Department team that has been guided by Jim Draughn since the '80s. With the assistance of Rudy Draughn, Gene Draughn, Tim Sawyers, Junior Sexton, Greg Ramey, Rosa Marshall, and Guynor Cave, Pike's supply inventory is kept at the ready. Ninety-five percent of whatever is needed by Pike's personnel anywhere is shipped to its destination by this department within two days of the request, and sooner, if needed. Routinely, supplies are shipped the most efficient and economical way. "However, hurricanes and ice storms affect the way we deliver tools and supplies dramatically," Jim Draughn said. "Almost nothing is

*Ashville, North Carolina.*

shipped in those demanding times." He explained that mechanics, supply room personnel, special runners, and garage employees are dispatched to be sure that crews quickly have what they need for storm work. "No crew member will work in an ice storm needing a sweatshirt, gloves or winter liners," Draughn assured. The supply room professionally supports the people in the field.

Possibly not known to the general public, but well known in the utility industry, is the contribution that Pike's Supply Department continues to make in product advice, modification, and development. Guynor Cave has an established reputation for tool-repair work that is often termed "amazing," and he is frequently consulted by manufacturers for his recommendations concerning tool improvements. The flame retardant vest, rope modifications, sleeves, hoists, and harnesses — these are only a few examples of the equipment that has been made safer as a result of the work and dedication found in Pike's Supply Department.

The build up of the '80s created more demands on this department as an ever-increasing inventory of supplies was required. When a need for more space became obvious, an existing building in Mount Airy, located on North Carolina Highway 52, was purchased. Called "Pike West," this is the department's present location and warehouse, and this facility accommodates the laboratory for rubber-goods testing .

The test-lab was under the able direction of Ben Brannock from its beginning in 1964 until his until his retirement in 1995. In discussing the lab's origin, Brannock said that he walked into the main office one day in '63 and encountered Floyd Pike, Lawrence Allen, Haywood Merritt, and Zack Blackmon, Sr., in conference. Allen asked Brannock if would be willing to run a test lab, if the company had one. Brannock said that before he could ask what a test lab was, Lawrence Allen declared,

*Mount Airy Supply and Test Lab.*

"We'll set one up, and Ben will run it!"

The decision was timely. The main office was undergoing a major renovation at the time, so that space was designated for the lab to adjoin Jim Draughn's small supply area, where he was assisted by George Wilhite and Delmar Ray Cook. Looking back over his years in the lab, Brannock said that he was assisted by many co-workers who were also his close friends. However, there was one five-year period that was a very special time, when he was assisted by Jack Cook, who had come in from the field following a heart attack. "This was a time of reunion for us," Brannock explained. "Jack and I had worked together for years on the line crew, and he was the lineman who had trained me and taught me everything that I knew about line work. He was a real professional."

The company is proud of the fact that Pike's testing requirements have always exceeded OSHA standards. In addition to testing all of the company's own gloves, sleeves, blankets, hoses, and bucket-liners, the lab has, from its beginning, provided testing services for municipalities, cooperatives, telephone companies, industrial customers, and even other contractors. No small feat for a company whose territory during this decade reached from Walter Jessup's crew working for the South Central Power Company in Lancaster, Ohio, to Bill Conner' s crew at the Lee County Co-op, in southwest Florida — approximately one-thousand miles apart. And the Commercial Wiring Division's work locations extended from the coast of North Carolina in the east to Texas in the west.

As Pike's territorial distance increased, personnel was once again faced with greater loss of time due to travel. In the crisis of the '70s, when numbers and the distance of jobs diminished, Pike's corporate airplane had been sold. By the late '80s, circumstances indicated the benefits that another plane would provide. A Beechcraft King-Air C-90 was purchased in 1988, and since then David Hall, chief pilot and co-pilot, Ken Shelton, have contributed to the efficiency of the company's overall operations safely and routinely.

# DISTANCE IS DISPELLED

Although Pike's service territory covers many states, distance is dispelled by the camaraderie and support shared by Pike's personnel. This fact is evident from the list of nicknames that had accumulated during the '80's, and continue to be heard even today.

Many employees will remember and identify their co-workers: "Abe"; "Airhead"; "Alligator"; "Barney Fife"; "Big E;"; "Bluegill Bob"; "Bo"; "Bracky"; "Boot Covers"; "Bulldog"; "Butch"; "Catfish"; "Chief"; "Chilly"; "Chubby";"Doc"; "Dooley"; "Droopy"; "Dub"; "Foreman J"; "Goughstein"; "High Pockets"; "Hillbilly"; "Horse"; "Hot Rod"; "Inspector 12"; "Inspector 13"; "J.J."; "John-E-Poo"; "Leroy"; "Liver Lips"; "Louie"; "Luke"; "Lum"; "Lurch"; "Mac"; "The Main Man"; "Midge"; "Monk"; "Moon"; "Mother Hen"; "Motor Mouth"; "M.R."; "Old Man"; "Old Man of the Mountain"; "Opie"; "Pee-Wee"; "Pop"; "Q-Berry"; "Red"; "Rose"; "Shade Tree"; "Sinkerhead"; "Slammer"; "Slick"; "Snodgrass"; "Snuffy"; "Sorry Rat"; "Stimey"; "Sue"; "Swooper"; "Tarpen Head"; "Uncle Milt"; "Wichita"; "Whitey"; and even "Zeus!"

*In the Pike organization, safety*
*has always come from the top down.*
*We've never been one for committees*
*that just make suggestions about safety.*
*Safety has always been one of the top priorities*
*of management. As safety director, my responsibility was*
*to keep everybody in this company from getting hurt!*

*—A.L. McAlexander*

# 32

# Impact of the '80s: Regulations and Technology

The 1980s were punctuated by governmental and technological developments. The support services within the Pike Company — especially the safety department— took a direct hit from governmental regulations, and operations throughout the company were impacted by the proliferation of personal computers.

When A.L. McAlexander retired on July 31, 1987, following thirty-seven years of service, Zack Blackmon, Jr., became Safety Director. And with the retirement of Emmett Forrest in 1990 — after forty-one years of service — the responsibilities of dealing with property damages, insurance, personal injuries and workers' compensation were also transferred to the Safety Department. With these changes, the role of the department widened so that it was not only responsible for employee safety, but loss-control as well.

With computers readily available, statistics could now be used to indicate where the emphasis should be placed in safety instruction, and statistics provided documentation that is computerized and maintained — a procedure which has become increasingly important. The Safety Department continues to be in charge of accident investigation. The department is also responsible for providing information and enforcement of all regulations, integrating safety throughout the company.

"Regulations—regulations—regulations!" said Zack Blackmon, Jr. "We deal with them daily." Although Pike Electric's safety rules already went beyond the new rules and regulations passed by the federal government in the late '80s, nevertheless, those changes resulted in an update of Pike's 80-page-manual to a 177-page safety and training manual, completed in 1991. Blackmon explained that government regulations added new rules and other information that pertained to right-of-way clearing, substation construction, commercial and industrial wiring, and the latest requirements related to overhead and underground work .

In the '80's, OSHA established the Hazard Communication Program, which required an employer to provide a separate booklet of information listing hazardous chemicals in the workplace. This costly requirement was put into force despite the fact that "to my knowledge," Blackmon said, "Pike has never had a serious hazardous chemical injury or accident." He emphasized that the Pike company has spent literally tens of thousands of dollars training its employees in the hazards of the chemicals around them, for example, gasoline, brake fluid, PVC glue, etc. "It is my contention," Blackmon said, "that our supervisors' time could be better spent doing line-work training. However, the Hazard Communication Program seems to be one of OSHA's pet peeves, and they cite and fine employers if employees are not trained. So, the training continues."

It is a general consensus among Pike employees, that the overall purpose of OSHA is good: to create a safe workplace for employees to work. "The problem with OSHA, like many areas of government," Blackmon explained, "is that you're not dealing with the individual who made the speech. Instead, companies have to deal with individual inspectors that seem ready to impose fines that are, more often that not, difficult to justify."

Blackmon said that the Pike company has always tried to stand for what it thinks is right. "Our reputation for being the safest in the industry was recognized in one of our appeals against an OSHA citation," Blackmon said. "The South Carolina Commissioner of Labor ruled that OSHA erred in its case against the company, which was based on Pike's failing to have an OSHA poster visible at the job site, no instruction sheet with the PVC-glue container, and no instruction book in a crew man's pocket — among other seemingly trivial complaints.

"At the end of the day," said Donald Anderson, who was present at that particular OSHA hearing, "the OSHA judge made the comment that Pike kept the best safety records of any company he had ever seen, and that there was no doubt in his mind that this was the safest company that he had ever had any dealings with."

"One of the most laudable qualities about Pike," said Attorney Guy F. Driver, Jr., who represented Pike at the hearing, "is the company's commitment to vigorously fight battles involving legal issues when the company believes itself to be in the right. In a world where many believe it is less expensive and more expedient to settle legal disputes, Pike makes no compromise on its principals or its integrity."

Following the hearing, the South Carolina Commissioner of Labor wrote the company a letter of apology for the citation."The South Carolina Department of Labor even sent a check to help defray attorney fees!" exclaimed Blackmon. "This is a rarity with OSHA." He framed that letter.

## PERSONAL COMPUTERS PROLIFERATE

Record keeping, statistics, inventories, fleet tracking, storm work — practically every task associated with the company's operation — was modernized in the 1980s with the introduction of personal computers and an upgraded computer system. "I remember when the first personal computer was brought in and put in the library for everyone to use," said Sandy King, executive secretary in the main office. "I spent every spare minute I could find playing with it and trying to learn how to use it." Even as late as '82, when Sandy King came to work at Pike in the payroll department, all invoices were hand-typed and corrections were made with white-out. "Imagine all the overflowing trash cans we had beside our typewriters!"

On-the-job training was provided then, and it continues today, as technology and soft wear change. "Actually, Tommy Jarrell walked and talked me, and many others, through Lotus! Barny Ratliff continues to help me still, when there is a problem." said King. "By the time I left the payroll department, everyone had a computer. It was just wonderful. Things have come such a long way." Amelia Reeves, executive secretary in the home office was away on leave, until her daughters entered school. "When I returned, I almost had to go back to school myself," she said. "Things were so modern around the office by then."

*Sandy King Executive Secretary in main office in Mount Airy at the time she began working with Pike.*

As the company identified more and more computer applications, the demands on Joe Reid's Information Systems Department increased so rapidly that by the mid-'80s, Coby Gough joined Pike as a programmer, and Kim Turney moved into programming as well. Later, the department converted to an IBM System/38, moving from a flat-file to a database system. To accommodate the system, an area in the main office was redesigned with an enclosed clean room, an uninterrupted power supply, raised flooring, and a halon fire-extinguishing system.

Integration of the company's computer facilities expedited the increasing accounting that is associated with taxes, and the profit-sharing changes which John Mullins deals with — responsibilites that differ in the various states of the company's service area. And seemingly, almost overnight, the tracking of personnel and equipment during storm trouble moved from an antiquated, albeit reliable, card file that had worked from the company's beginning, to a new era of immediacy and efficiency.

*Storm restoration is a Pike specialty.*
*Snow storms and ice storms, hurricanes, tornadoes —*
*whatever the natural calamity,*
*Pike crews get the power back on quick.*
*They are experts!*
*Pike does more emergency work*
*than anyone else in the business!*

*—Floyd S. Pike Electrical Contractor, Inc.*

33

# Storms of the '80s

For the people of Pike, storms are an integral part of the memories of the 1980s. When ice came into the company's service area in December, 1981, the restoration work continued well into the new year. The following December was even worse. The *Pike Line* described the conditions then as "some of the worst weather the South had seen in half a century." The Carolinas, Georgia, and Alabama suffered major power outages.

Pike crews were called out early on New Year's Eve to restore power to Duke Power customers in western North Carolina. Their rapid deployment and the dedication of Pike's people were gratefully appreciated. "During the recent storm emergency," wrote Arnold Monteith, Duke Power employee at Pisgah Forrest, North Carolina, "your foreman Elwood Dowdy exemplified what a foreman should be in his evaluating the situation and proceeding in a very expeditious way, getting all the repair done, and, at all times, keeping in mind the safety and well being of us all. Lineman Mitchell House was very impressive in that no hesitation was shown about climbing under the very adverse conditions encountered, and his helper, Charles F. "Buddy" West, seemed to have had schooling in public relations when checking with customers." Monteith concluded that this crew was "the best I have been privileged to work with during storm emergencies."

In South Carolina, Georgia, and Alabama, winter stuck another blow with more ice only a few days later. Pike crews were called out again and sent to various work locations throughout the South. Most of the North Carolina crews that normally worked for Duke Power Company in Hickory, Salisbury, Durham, Spindale, Winston-Salem and Charlotte were sent to Alabama Power Company for restoration work in the Birmingham area where more than 750,000 customers were without power.

Further assistance in Georgia and Alabama was provided by Pike crews from as far away as Fort Myers, Abingdon, Virginia, and Kingsport, Tennessee, who traveled to assist Coweta-Fayette Co-op in Newnan, Georgia. Pike's crews from Grayson, Kentucky and Rainsville, Alabama went to Jackson, Georgia to restore power for Central Georgia EMC.

During the storm emergency, Pike crews traveled 60,000 vehicle miles in the first twenty-four hours with no major equipment problems. Pike's field mechanics from North Carolina made the trip to Alabama, while the field mechanics from the Georgia garage carried out the equipment maintainence service that was required for their own hard-hit area.

Following the storm, one of Alabama Power Company's customers issued to Pike a compliment and a request:

"I really appreciate the great job your personnel did. And if we ever have another ice storm, I hope your people will be down here again. Your men are workers and know what they are doing." The customer's wish was granted. The Pike company has continued to work for Alabama Power Company.

If trouble comes in three's, the third December storm in as many years hit the South in '82, with severity that had not been predicted. Little did Pike realize that the early calls for help were but a warm-up for what was to come in the next thirty-three days.

By 3:30 a.m., December 12, Pike's crews in the Pulaski, Abingdon, Bluefield, Hazard, Pikeville, and the Prestonsburg divisions of American Electric Power were the first at work. As wet snow continued to fall, crews were requested by AEP to go to additional locations in Virginia and West Virginia. Crews based in North Carolina — from Dobson, Waynesville, Mount Airy, Eden, Spindale, Hickory, Salisbury, Durham, and North Wilkesboro — went to Pulaski, Abingdon, and Fieldale, Virginia. Seventy-two additional men and thirty-seven pieces of equipment were soon sent to help in what turned out to be a major storm.

Only thirty-three days later, Bill Conner's crews in Fort Myers, Florida, were put on standby at their location. The EMC's in Georgia were watching the weather forecast and were requesting additional help. Half of the crews in Florida were flown to the Pike office in Monroe, Georgia, to get equipment that was ready and waiting for them there.

As the ice storm tore through south-central Georgia, Pike crews were already hard at work, even though the magnitude of this storm had not yet been realized. Within a few days, all of the Pike crews in Georgia were involved in storm restoration. Before it was over, there were so many requests for help that all but sixty-one of the total people in Pike's power line division were involved in the effort. Bill Conner's crew traveled the most miles of any one crew to work — over 1,400. 30,000 man-hours and 22,477 hours of equipment time were logged with no major personal or vehicular accident.

In the '80s, Pike crews in Georgia and Florida handled Hurricanes Kate and Elena where damage to the Florida panhandle was severe but confined. "Hurricane Kate delivered a direct blow to our area," Talquin Electrical Cooperative, Inc., confirmed, "totally destroying some of our electrical system and severely damaging much more." Pike's storm restoration was consistent with its reputation. "As was the case during Hurricane Elena, the productive, courteous and safe manner in which your employees assisted us, has gained the respect of all of us associated with Talquin Electric, and will be long remembered."

Before the decade ended, Duke Power's service area received astonishing damage in the summer of 1989. From May 5 through May 11, severe thunderstorms and tornadoes caused approximately 250,000 Duke Power Company customers to be without power at one time or another: 207,000 in North Carolina, and the remaining 43,000 in South Carolina.

Pike responded quickly to Duke's request for help by moving over forty crews from the American Electric Power System into Winston-Salem early Saturday morning, May 6, where over half of the division's circuits were out and thirteen of the major transmission towers had been toppled.

To assist with damage in other parts of the Carolinas, an additional fifty crews were mobilized in the early morning hours that same day from Georgia Power Company and from various EMCs in northern Georgia. The Georgia crews worked first in South Carolina, and then moved on to Hendersonville and Rutherfordton. Finally, they too, were needed in Winston-Salem where the restoration required "thirty-four tractor-trailer loads of materials, eighteen loads of poles, and a large number transformers," according to Paul Briggs, Duke Power vice-president of the Winston-Salem division at the time. He added that the power company sent about five tractor-trailer loads of scrap to its warehouse, and that about 1,000 poles had been replaced when the restoration was completed. During the height of the storm, Pike had over 1,000 people and 400 pieces of equipment involved with the repairs. Pike's garage forces, warehouse personnel and office staff worked extraordinarily long hours.

Paul Briggs said, "Duke Power Company has never seen anything like this storm!" That was in the spring.

Hurricane Hugo came in the fall!

*Pike's restoration in Winston-Salem, North Carolina.*

*I've been on many storm calls.*
*Often, when I leave home,*
*there is no power there,*
*and I'm heading down the road*
*to restore power somewhere else.*
*It makes you when feel kinda bad*
*to leave your own family without power,*
*when you are going to put somebody else's power on.*
*But I remember when I was small,*
*and we were without power at home,*
*my father, James Anderson,*
*left us many times to go to work on storm trouble.*

*—Donald Anderson*

34

# HURRICANE HUGO

Not since Hurricane Hazel, in 1954, had the Carolinas felt the effects of a hurricane statewide. Long before daylight, on September 22, 1989, Pike's main office was getting its own telephone lines up and going, and storm logistics were already in progress throughout the organization. As calls for help came in from all parts of the company's service area, the entire Pike operation was engaged. The garage and Supply Department were well prepared. Safety supervisors were mobilized. Within a very few hours, hundreds of crews, well-trained, drug-free, and prepared to work in incredible devastation, danger, and deplorable conditions were en route to destinations that covered a wider path of destruction in the Carolinas than any hurricane's before or since. Many of the crew members were leaving their families without power in their own homes — some even with considerable damage — to go to areas where homes were in ruin and residents were even more distressed. Akers Woods and Charles Easter dispatched the field mechanics to every location where equipment needed for repair or replacement — even as far inland as the Virginias.

"Hugo cut a path of destruction directly through our main service area from Charleston, South Carolina, to just outside Charleston, West Virginia, putting all of our people in the affected area to work," said Joe Pike. "We also mobilized additional crews from as far away as Ohio and western Kentucky, and from Fort Myers and West Palm Beach, to assist in restoring power." Raymond Dickson's crew actually rode out the storm in Charleston, South Carolina, and started to work as soon as the winds died down. At the peak of restoration, Pike was simultaneously working for four power companies, nine co-ops and three municipal systems, with as many as eleven separate operations on one customer's system.

"This was certainly a test of our storm preparedness," Joe Pike said, "a test, I feel, we passed with flying colors." He said that the successful operation was the result of the "tremendous effort and total team work of everybody in the Pike organization." More than 1,400 people worked in the mud, rain, and timber debris. Pike had more than thirty supervisors in the field, and seventeen field mechanics were responsible for keeping the equipment running. The garage and supply room personnel kept the supplies stocked and moving. The office force and management staff manned the storm center. Primary restoration required four weeks and 485,000 man-hours. The worst injury during the entire ordeal was one broken foot!

Many residents provided some relief for the men on the crews as they worked in deplorable conditions. Pike crews will never forget the manager of a Hampton Inn in Charleston, South Carolina whose brand new facility was almost ready to open when Hugo hit, flooding motel's the ground floor.

Nevertheless, Pike's men were welcomed to occupy the remaining quarters, and the manager turned the entire motel over to Pike for an operations center, a radio-communication site, and, equally important, a place to sleep.

Initially, Pike's own personnel transported food for hundreds of miles to their co-workers. The work sites in the worst-hit areas were virtually paralyzed when the first crews arrived, and for many days afterward. Somewhere in the Charleston, South Carolina area, however, a shrimper may still be at work who had a boat with a generator. He somehow managed to find a dockside café that was still standing after the storm, and with the help of the generator, hot food was prepared for Pike's personnel. For other crews, who were working far away from the few restaurants with power or lesser damage, hundreds of persons in neighborhoods and outlying areas showed their appreciation by sharing whatever food they had with Pike's personnel, as power was restored.

Pike crews were among the first to begin work and among the last to be released in every area of the storm's devastation. "And we had over one hundred employees who worked down there for probably a year and a half on cleanup work returning the systems in South Carolina to their permanent positions," said Ralph Anderson. "This is the highest compliment that can be paid in regard to our customer's satisfaction with our performance," said Joe Pike. "I don't think anyone could be prouder than I of the tremendous work performed by everyone." Pike added: "Everyone on the Pike team should be proud of our company's performance."

Similar words had been said about that first crew pulling wire many years before: "Pike performed!" And Pike was continuing to do so, quite remarkably forty-four years later, under the worst possible conditions.

**Rutherford Electric Membership Corporation**

202 HUDLOW ROAD • FOREST CITY, NORTH CAROLINA 28043 • TELEPHONE (704) 245-1621

October 11, 1989

Paul Greenhill
Floyd S. Pike Electrical Contractors
P.O. Box 868
Mount Airy, NC 27030

Dear Paul:

We, at Rutherford EMC, want to thank you for the help given to us during the aftermath of Hurricane Hugo. Your crews did a splendid job, worked very hard and expertly with Rutherford EMC personnel, and were a great help to us restoring service. Please extend our thanks and appreciation to each of them.

Sincerely,

Gary

Gary A. Whitener
Manager

**NORTH**

October 9, 1989

Mr. Floyd S. Pike
Pike Equipment & Supply Company
P.O. Box 868
Mt. Airy, NC 27030

Attn: Mr. "Buddy" Pike, President

Dear Mr. Pike:

I once again had the opportunity to meet some of your linemen as they were in Charleston, S.C. restoring our power. I am always so proud to say I know and work with the good people at Pike Equipment. You should know that your employees stand heads above all the other contractors I have had the pleasure of meeting during my 20 years in this industry.

Your personnel are proud, take pride in their work and even look more professional than the rest of the crowd. We at North Hand Protection are honored to be your supplier of protective equipment and we look to many more years of continuing service to you and your fine organization.

Please pass the word to your crews that we will do whatever we can to help them in any way we possibly can. All they have to do is call us at 803/554-0660, we'd be honored to help.

Best regards,

Max E. Klemm
Sales Manager
North Hand Protection

MEK/ebd

cc: TJKulp
JKHane
RRHicks

North Hand Protection
A Division of Siebe North, Inc.

4090 Azalea Drive
P.O. Box 70729
Charleston, SC 29405
803-554-0660
Telex: 4611157

**BERKELEY**
**ELECTRIC COOPERATIVE** INCORPORATED

P. O. Box 1234 • Moncks Corner, South Carolina 29461 • (803) 761-8200

October 19, 1989

Mr. Zack Blackmon, Sr.
Floyd S. Pike
P.O. Box 868
Mt. Airy, North Carolina 27030

Dear Zack:

On behalf of the consumers and employees of Berkeley Electric Cooperative, Inc., I extend my sincere thanks for your help during the long recovery process after Hurricane Hugo.

Your personnel should be commended for their hard work. I believe that Bobby Mayfield has been here since the day after storm. We appreciate his dedication. All of your crews have represented you, their families and their hometowns well. Thanks to their hard work, our original estimate of six to eight weeks for complete restoration was shortened to three to four weeks.

Again, thank you for your help.

Sincerely,

Skip

E. E. Strickland, Jr.
Executive Vice President
and General Manager

*More than 1400 Pike employees worked in Hugo's mud, rain, and devastation.*

*Enroute to "Hugo", a Pike caravan stops for a meal.*

*When Pike says its people are its greatest asset,*
*it has the numbers to back it up!*
*The company's twenty-six field supervisors*
*are all promoted from within,*
*and have a combined 557 years of service —*
*a 21.4 year-average.*
*When we get them all together,*
*nobody else comes close!*
*—Floyd S. Pike Electrical Contractor, Inc. (1990)*

35

# The Pike Supervisor: Key to the Operation's Success

"Although their duties have changed over the years, some things are still the same," said Joe Pike. "The heart of the company is the supervisor, and to many of our customers, a Pike supervisor is 'the company.' " Pike Electric has never had a "Pike salesman" or a sales department, but everyone is a salesman to a certain extent. "A Pike supervisor is responsible for his area," Joe Pike explained, "and when he 'sells' to a customer, he is the one to deliver on the promises that he makes to that customer. To do this from day to day is a continuing achievement," Joe Pike observed. "Under storms conditions, it is quite a feat!"

Where is a crew to work? Where to set up a operation center? Where to find food and gas at three o'clock in the morning in a town with no power, where road signs are blown down; and after hours in the rain and mud, where to get a shower, dry clothes, and some sleep? These are only some of the additional responsibilities of a Pike supervisor when his crews are mobilized.

"After Hugo hit, Jim Benfield and I pulled our tour of duty in the storm center where we were coordinating and dispatching crews; and then, we were sent to Moncks Corner, South Carolina for a week or so," said Jeff Collins. "I've never seen a place so devastated. I've never seen as many yellow trucks at one place at one a time, with so many of Pike's people restoring power. It was an awesome sight to see the way our supervisors were able to handle such a massive amount of crew activity. That made a very big impression on me, as I was coming along with the company."

The routine responsibilities of a supervisor, if not as exciting as those during a storm, are at least as challenging. "The job of an area supervisor is probably the toughest job in our company on a day-to-day basis," said John Merritt, vice-president, Southern Division. "The hours and dedication required are not something that you can adequately explain." Merritt said that he always enjoyed his time as a supervisor, "even though it was a tremendous challenge." A supervisor has a mixture of responsibility. "He has the enjoyable charge of keeping close ties with the crews and their work, as well as the close contact with a customer," said Merritt. "However, he is also responsible for anything that goes wrong — for example, property damages; all the while he is trying to prevent accidents of any kind. I remember every night waiting for the phone to ring, and, seldom was the news good." Merritt explained that government regulations consume so much of the supervisor's attention today, that a great deal of the time that once was devoted to crew contact, has been passed to the general foreman.

"Some crew members have the perception that it must be nice to be a supervisor, to sit in a pickup truck and ride around," said Jeff Collins, "but the supervisors have an awesome responsibility just in hiring people — the right people — those who can produce good quality work for a customer, who can work drug-free, and in a safe manner."

Supervisors are responsible for training foremen and employees as they progress. "They are, in essence, almost a father-figure to our people out in the field," Jeff Collins observed. "This is not a job from 7:00 in the morning to 5:00 at night. It covers everything about a job — from taking care of truck break down the very first thing in the morning — to the hours spent making work plans, and more hours spent calling people from home late at night to implement the tasks." Jeff Collins added, "I can't say enough about how I deeply and truly respect the work that our supervisors do for this company."

"I know that we, in management, won't forget the contribution that our supervisors have given to our company and the loyalty that they have shown through all their years of service," said Claude Brinkley. "I'm proud that today's management had teachers like Posey Edwards, Jack Farris, Richard Simmons, Leo Sams, Fred Anderson, Frank Jessup, and so many others outside of my division, who had the ability, the knowledge, and the guts, to stick with it, and teach the people that we have today to do the kind of work that we do."

Much of Pike Electric's success can be attributed to its supervision. "Sometimes when we have a supervisor's meeting," said Ralph Anderson, " I just look around — there is so much talent and so much experience in that room. I don't know of any other company that could even compete with it."

*Supervisors conduct training on site*

*Area supervisor Phillip Combs with crew in Georgia*

*Early Supervisors Mtg. Blue Ridge Hotel, Mount Airy, NC*

*Supervisor Richard Simmons on the job.*

*Long-time supervisor Jack Farris.*

*Joe Pike presents Fred Anderson, long-time supervisor, plaque of appreciation at his retirement.*

*I suppose that everyone has certain moments in life*
*that are indelibly fixed in their minds.*
*I can remember one such moment*
*just before Christmas in '78 as if it were yesterday.*
*Our family was traveling from Lenoir, North Carolina,*
*where I then worked for Duke Power, to Narrows, Virginia*
*to visit our parents for the holidays.*
*As we approached the Mount Airy city-limits,*
*I asked Marsha what she would think about living there.*
*"Are we being transferred?" she asked.*
*When I said, "No," the car became very quiet.*

*— Barny Ratliff*

# 36

# The Supervisor's No. 1 "Employee"

"A supervisor's wife has to be his No. 1 'employee', " said Roy Thomas, "because she's always required to answer the phone, take notes and telephone numbers in order to keep track of her husband's employees and relay their needs." "Cookie" Jessup, wife of Frank Jessup, area supervisor in Ohio, verified Thomas's statement. "I don't know what Frank did 'b.c.' — before 'Cookie'," she said. "I have taken care of his calls for almost thirty years now."

The nature and the ever-changing location of the company's work, along with the sometimes unpredictable work schedules, demand flexibility, competence, and understanding from the wives and families of the men at Pike. Many times, the wives of Pike's personnel have even moved their families — more than once, like Mary Thomas — to new homes in new towns when a job continues for quite a while. Routinely, Pike's wives must manage the responsibilities of a home and a family without the day-to-day help from their husbands — and often for long periods of time, especially when the location of a job is far away.

When Jack Farris retired as area supervisor in 1991, his wife, Mamie, said that this was her retirement as well. She said that in the 44 years of her husband's employment, she had never failed to make his breakfast, even at 3:00 a.m., if his departure required it; and she had made dinner at midnight, if that was when he was finally got home. Mamie Farris is not the only one. Katherine Ashburn did the same for 36 years; and, in addition, the gourmet lunches she packed each day for her husband, Everett "Ted" Ashburn, often left the other men on his crew wanting, especially, when their noon time fare was frequently no more than a quick snack from a convenience store.

Distance and long periods of separation go with a lineman's job. Leo Sams' daughter, Leslie, said that while growing up she and her brother Mark, now a Pike area superintendent, Central Division and third-generation employee, always called themselves Pike "orphans". It turned out to be good training for becoming what she called a Pike "widow," after she married Mark McPeak, an "A" lineman in Lexington. Her mother, Lottie Sams, set the example, because she too, called herself a Pike "orphan" and a company "widow." Lottie Sams' father, Lester Inman, was a lineman who climbed with Floyd Pike at Cherry Point, and Inman joined the company in 1946. Daughter, Lottie, knew exactly what to expect when she married Leo Sams.

"Bring your wife to the interview, if possible," was the directive given to Barny Ratliff many years ago. "I had interviewed for jobs at several companies prior to my college graduation," said Ratliff, "but I was unprepared for this one request that I got from Pike. While it seemed rather unusual at the time, I have come to understand the unique role that wives play in the operation of Pike Electric."

The demands of this unique role were brought home to Kelly Merritt, wife of John Merritt, during the disorder caused by Hurricane Hugo. Kelly Merritt said that when the storm hit in Mount Airy, trees were crashing all around, electricity was already gone, and the whole place seemed ready to blow away in the wind when John Merritt left the house in the middle of the night. "He said that he would call when he could," Kelly Merritt said, "but only 'in case of death' was I to call him, with all the work that is involved in crew mobilization." She was terrified and immediately called her mother in Charlotte, whose comfort was understandably nil. "Mother said that she had a tree through the roof, and rain was pouring in everywhere, so she didn't have time to listen to my worries about being alone. I knew that from then on, that I was on my own."

Incidentally, Kelly Merritt said that when a Pike crew pulled up to her parents' house to restore power many days later, her mother was so happy to see them that she ran outside to greet them. She asked if the crew happened to know her son-in-law? Indeed they did. Marvin Hill and his crew were there to assist, and they had worked for a time directly with John Merritt, when he was their supervisor.

Opal Pike established the tradition of support that the wives of the Pike management team have carried on for decades — whether by bringing food to the storm room, making meals at all hours of the day and night, attending ball games and school plays alone, or juggling a job, home, and a family to accompany a husband in his responsibilities.

Not to mention the hours of waiting and worrying known to all of these women. It is often not an easy life for any of Pike's families, especially in times of crises that come to every household — and, which always seem to occur, when *he* is away. However, there is a common bond of understanding and support among the wives of the men at Pike, that results in the camaraderie of a unique and caring, extended family. Shirley Brinkley, wife of Claude Brinkley, spoke for many, when she said, " Pike is, and has always been, my family."

*If you like hard work and the outdoors,*
*if you're tough and smart, then Pike Electric wants you.*
*You'll start as a member of one of our crews*
*and learn on the job.*
*As the new guy on the crew, you'll start out doing*
*the hardest, roughest and dirtiest work.*
*Every supervisor and executive at Pike worked his way up,*
*and you can, too.*
*They know how to help you and how to get out of your way.*
*If you've got what it takes to prove yourself on a Pike crew,*
*then you're good. You can handle just about anything.*
*If you're a hard worker, who takes pride in what you do,*
*if you can think for yourself, and work with people,*
*then you have a real future at Pike.*

*—Floyd S. Pike Electrical Contractor, Inc.*

# 37

# A New Team

Toward the end of the 1980s and into the early 1990s, Pike Electric experienced a period of major transition. Claude Taylor, assistant vice-president, Northern Division, died in 1984, after twenty years of service with the Pike company. For most of his career, Claude Taylor worked on the American Electric Power system in the Abingdon, Virginia area. He had been hired and trained by Shorty Clawson. In turn, Taylor trained numbers of Pike's supervisory personnel.

Beginning with the retirement of A.L. McAlexander, many other Pike supervisory and executive personnel concluded careers that had lasted for decades.

With the retirements of Haywood Merritt, Jr., Emmett Forrest, Jr., Zack Blackmon, Sr., Roy Thomas, Claude Brinkley, Glen Pike, and Glenn Dobson, a large part of Pike's executive management team that had been in place, for the most part, since the '40s and '50s, passed the responsibilities of their offices on to others.

*Claude Taylor*

Barny Ratliff assumed the duties of executive vice-president when Haywood Merritt retired October 25, 1991 after thirty-eight years of service. At the time, Ratliff had been out of college for a little less than five years and had worked for Duke Power Company for the entire time. "I had held four positions, lived in three different cities, and was the youngest superintendent of engineering, construction and operations on the Duke system. I was comfortable with my future and felt very safe," Ratliff said. "However, something was missing."

While having lunch with Ralph Anderson, who was supervisor in the Lenoir area at the time, he learned that there might be an opportunity with Pike. "I'm not sure that I was being recruited, but to this day Marsha credits Ralph with our move to Pike. I suppose I do also," Ratliff observed.

Ratliff had long been familiar with the company, even prior to his work with Duke, because his father, Robert Ratliff, Sr., had had frequent contact with Pike during the many years that he was in charge of Appalachian Power Company's Glen Lyn Steam Plant at Glen Lyn, Virginia. "The future of the Pike company and my personal future became totally connected when I decided to give Ralph Anderson a call," Barny Ratliff said. "The feeling of teamwork and pride that was exhibited by the Pike organization had drawn me to the company."

Claude Brinkley retired June 30, 1995 after thirty years with Pike, and Jeff Collins became vice president of Pike's Northern Division. "In '84, after I had mailed a resume to Pike, Claude Brinkley called one Thursday night about seven o'clock and wanted to know if I would be interested in an interview," said Collins. "He told me to come at three o' clock on a Saturday afternoon for what turned out to be only the first interview —

but certainly not the first surprise. I was expecting a 'coat-and-tie' type of meeting, but Claude was wearing blue jeans, boots, and a flannel shirt!" Collins said that he left that meeting with many doubts about being hired, especially when Brinkley told him that there were seven other applicants applying for the job. His hopes were revived, however, when he was called to a second interview, this time with Claude Brinkley and Haywood Merritt. "By then, I was getting very, very excited about coming to work. I hadn't really known much about Pike — had no idea how big the company was — had no idea it covered so much territory. But after the first interview, I started asking questions!" Finally, he was called a third time. This time he met with Brinkley, Merritt, and Joe Pike. "After this interview, I asked Claude how things looked for me, but, again, he didn't have a whole lot to say, other than that he would let me know." Collins said that another surprise occurred when Claude Brinkley called on a Sunday night and wanted to know when he would be ready to start to work. "I was thrilled to death. I told him that I could start the next morning, but he said he really didn't need me quite that soon." However, Collins said that he did begin the following Monday, with Jim Bingman' s line crew in Bland, Virginia.

"It was a good learning experience to work in the field. In fact, there is no way to do the job that I have now, unless you have actually had the field experience to know what goes on. You wouldn't have a clue if you didn't. I saw for myself that everybody out there worked hard and was interested in doing a good job. They had to — they were on units — where so much work must be done in a day's time in order to make money. Then and there, I learned that the team concept is a must, if a line crew is going to be both safe and productive." Collins said that, for a period of time, he was then shifted among many crews, working as the fill-in person; and although this roving allowed him to gain experience, he was always the groundman. "I dug a lot of rock holes and shot a lot of Kinepak."

In discussing the progress of his training, Jeff Collins said that after he left the field, he circulated among the supervisors in the Northern Division to help out in any way that he was needed. "Then sometime later, Claude and I were sitting down, scratching our heads, trying to decide who the foreman should be on a job that we were just starting. Finally, he said, 'I'll tell you what, you just go do the job.' "

Jeff Collins then spent the winter in Bluefield on a duct-bank job. "It was a really good learning experience for me," Collins observed. "Being a foreman, running a crew, having the headaches and the problems, and making money too — the job was production-based and still had to make money. It was like being thrown into a river to sink or swim. But I really did enjoy it."

Collins added that another good thing about his training among many crews on a variety of jobs was that he worked closely with almost all of the foremen in the Northern Division. "The benefit of that carried over to my responsibilities today, because I know my foremen, and I know their capabilities, their mannerisms, and their personalities," he said. "This allows me to know which foreman would be the best for the job in any situation that arises."

When Roy Thomas retired in December 1994, after forty-seven years of service, his responsibilities as vice president of the Southern Division were taken over by John Merritt, the son of Haywood Merritt, Jr. "I have been with the company, literally, all of my life," Merritt said. "I grew up in the company. I remember flying in the right seat of the company's plane with Maynard Snow while going with Dad to look at work in West Virginia. I remember riding with Bobby Cox to Abingdon to deliver equipment, and, when I turned sixteen, Dad put me to work." Merritt said that from then on, he worked every summer except one, every Christmas, and any other break from school, until his college graduation in 1984. "I never knew what I wanted to do after college," he said. "I did know that a job that confined me to a desk or the indoors was not for me, but I really wasn't sure what direction was right for me and truly had not considered working for the company before my last semester of school. It did not take long to recognize that I had made a good decision!"

Like Jeff Collins, John Merritt's first year was "quite memorable" in several ways. "My very first day on the job was in Columbia South Carolina. I drove down on Sunday to make sure I could find the show-up (location). When Monday morning came, I managed to lock my keys in the trunk of my car in the motel parking lot. Needless to say, I was late and quite embarrassed — not to mention short of money — I had to pay a locksmith to open my car door. I hadn't made the first impression on the foreman that I had hoped I would."

John Merritt was also moved "from location to location" throughout his first year. "Only in retrospect," said John Merritt, "do I see the benefit of having worked in such diverse areas of the company." He said that there is nothing more important for management, than a thorough working knowledge of our industry. "We learned this the only way possible, by doing it. My only regret about my time in the field is that it was short-lived."

Training under Glenn Dobson, assistant vice-president who retired in 1990, after thirty-one years of service, had a strong influence on John Merritt's development at Pike. "Glenn Dobson was always very demanding. At the same time, he was fair and consistent. I admired Glenn for his desire to continue learning new things and educating others in the process. I never spent any appreciable amount of time with Glenn that I did not gain something. Glenn also forced everyone around him to conduct themselves in a manner that was one notch above par. Most importantly, he was always out in front setting the pace that we all hoped we could follow."

Glen Dobson's training also prepared others in his division for management positions. "When I began with Pike in '70," said Jerry Boatright, who was named assistant vice-president, Southern Division in 1994, "there were only a few hundred employees, and Glenn Dobson was the only supervisor n Georgia. He came by our work site on a regular basis with tools and supplies. He always made sure that our rubber gloves and sleeves were changed out according to company policy and that we were working in a safe manner. His training, along with that of Colbert Beasley and Roy Thomas — in my opinion, three of the company's best — have been responsible for the advancements that I have been able to make with this company."

Boatright explained that his responsibilities have grown enormously throughout the past twenty-five years. "I have held the positions of a groundman, lineman, foreman, area supervisor, and, today, I serve as an area manager and a company officer. I am responsible for Pike's operation in Georgia and Alabama." Boatright's career illustrates the fact that the Pike company promotes predominantly from within.

Training provides upward mobility throughout the organization. Todd Drane, an employee in the Mount Airy garage has been with the company for ten years. "I started to work, at first, helping out on the yards, then delivering equipment to job sites," he said. "With the training I have received, I am now a mechanic qualified to work on bucket trucks."

"This is one of the many benefits that this company offers," Jerry Boatright said. "I am very proud of the advancements I have been able to make within the company, promotions that are the result of hard work and good training."

Glenn Dobson also trained Zack Blackmon, Jr., who became vice-president of Human Resources in February,'95. "One of the first bids that I put together was the Hammond Crossing 115 KV Transmission Line at Cumming, Georgia," said Blackmon. "I'll never forget walking every step of that six-mile line and looking at every structure location with Glenn Dobson. He was very persistent." Blackmon explained that one of the lessons that was reinforced under Dobson's supervision was that a crew did everything possible to keep the work productive.

"We were working in Cumming, Georgia, and I had finished up my week early on Wednesday, in order to come back to Mount Airy for the weekend," said Blackmon. "I drove back to North Carolina, and when I got home about seven o'clock that evening, there was a message waiting for me — yes, from Glenn." Dobson explained in the call that the crew in Georgia badly needed a compression tool from the garage in Mount Airy by the next morning or the crew would be "shut down" Thursday and Friday. "I got the compression tool from the Supply Department," Zack Blackmon, Jr., said, "and headed back to Georgia. My crew was ready to go to work at 7:00 a.m.!"

Blackmon went on to say that a situation like this was quite familiar to him and his response was not surprising, since he is the son of long-time Pike executive, Zack Blackmon, Sr. "I was born into it, and I have grown up in the company. My first memories of Pike were seeing Zack, Sr. coming home from work at the end of the day with dirty, muddy clothes, and our family moving to Spindale when I was about four years old, so that Dad would not have to do so much traveling to be near the crews that Pike had working there. I remember coming to the main office in Mount Airy in the mid-'50s, when it was just a little white building with a garage."

Zack Blackmon, Jr., came into the company full-time in 1976, and his training continued under Lawrence Allen, Richard Simmons, Audie Simmons, Roy Thomas Colbert Beasley, and many others. "From these instructors," Blackmon said, "I learned early on that every one of us does whatever is necessary to get the job done." Whatever is necessary in today's world is extensive, and Zack Blackmon, Jr., is in charge of a department that deals with all aspects of Pike's employment.

"We have *people* that we work with here, not human resources to be managed like a commodity," said Joe Pike. "We have gone along with the terminology, 'human resources' to be compatible with practically all of the rest of the business world." With an overt dislike for the term, but excited about the new department's establishment and its potential, Joe Pike said, "I hope and I certainly, do not intend for our department to become what the term implies, but the department is a vital new addition to the organization that has been created to relieve Pike's supervisors of as much of their bureaucratic load as possible."

In addition to the safety responsibilities that continue, Zack Blackmon, Jr., and his staff are responsible for a wide range of concerns that includes everything from insurance to traffic-flagging certifications and everything involved with personnel, from new employment to direct deposit-banking, child-support garnishments, workers' compensation, monthly, random drug-testing, claims, licensing, and unemployment — the list seems endless.

The tremendous demands on the department have also impacted the responsibilities of Kelly Surratt. Before becoming assistant vice-president, Surratt was Pike's main office manager, and in that capacity, he had already initiated the processes of screening and requiring home-office approval of all applicants before they were hired, which was a great cost-savings to the company and also aided in providing the company with top-quality people. "This screening process has been improved over the years," commented Reg Banner. "It is still not an easy job, especially, when some of our supervisors are in desperate need to hire employees. Kelly has a way with people, however, and he can give them bad news without making them mad."

Kelly Surratt explained that now when a person is hired, a pre-employment physical and a drug test are required and references and driving records must be checked. Also, much of Surratt's time is required to interpret and implement the numerous laws that relate to hiring practices now. "The American with Disabilities Act, the Age Discrimination Act, the Pregnancy Discrimination Act, and various other laws that pertain to race and gender have been passed with good intention," said Surratt, "but they create a lot of red tape for companies in their aftermath."

The employment and requirements of people who are spread all over the southeast is a difficult process to monitor and to maintain. Having worked and trained in all of the company's divisions has really been essential to Zack Blackmon Jr.,'s success in establishing the human resources department and dealing with its many aspects.

"Getting to know Pike's people," said Zack Blackmon, Jr., "their diverse working conditions in many different areas, and working with McAlexander in Safety, as well as the time I spent on the job in the field — all have provided the background, knowledge, and experience that directly relate to my responsibilities."

Hard work and good training were essential to Eric Pike, son of Joe Pike, when he assumed the responsibilities of vice-president of the Central Division after Zack Blackmon, Sr., retired in 1994, following almost forty-seven years of service with Pike. Like John Merritt and Zack Blackmon, Jr., Eric Pike commented that it was hard for him to establish a real date for the beginning of his work with the company. "I have always been a part of Pike Electric. Some of my earliest memories include the times that Dad would take me over to the office to climb on the trucks. I remember sitting on Pa Pa's lap in his office and playing with the sparkler toy that he kept in his desk for me. From the time that I was two years old, I attended the Georgia EMC's managers' meeting at Jekyll Island, Georgia, every June, until I was old enough to work in the summers," said Pike. "It is really a pleasure now to get to see many of the managers in a business setting that I met growing up and have known practically all my life. There were always customers, suppliers, and investment people visiting in our home. This company has always been, and continues to be, a central theme in my life."

Eric Pike said that when he was sixteen, he needed a summer job, "and this was the easiest place for me to get one." He was first put to work as a helper in the Commercial Wiring Division with a crew that was wiring a new stone-cutting building for the North Carolina Granite Quarry in Mount Airy. "When that job ended, my foreman, Delmar Lewis, and I were then sent to work on a Hardees fast-food restaurant in Greensboro. We would leave each day at 6:00 a.m., and we would get back about 6:00 that evening — if everything went well. That was my first experience with long hours. It wouldn't be the last."

Before that first summer was over, Pike knew that he loved construction work, and although he didn't know a lot about being an electrician, he knew that he wanted to know more. When the next summer rolled around, he again asked his dad for an electrician's job, but Joe Pike suggested that since Eric Pike was old enough to work on a line crew, he should. "Dad really thought that I would like it, but I wanted to go back to Commercial Wiring because there was still a lot there for me to learn." Eric Pike said that his dad made a deal with him. "He said for me to try working on a line crew for two weeks, and after that, if I still wanted to return to commercial work, I could."

"During the time," said Donald Anderson, Central Division area supervisor, "I was foreman on a job that we had near Dobson with the Surry-Yadkin EMC system. One day our area supervisor Jack Farris, came over to tell me that he was bringing a new man to work with me during the summer months. This was nothing unusual because Jack was always bringing new people to work in the summer — school people and all. He said this one was going to be a little special, because he was the boss man's son, Eric. I always knew Eric was around, but I never did expect him to come out and work with us, at least, not where I was working. We would just see how long he would last out there! We were on unit-work, and we worked hard. The type of work that we were doing was reconductoring — hard work — line work that involved a lot of climbing. Monday morning, Eric came walking in, and he was ready to go when he got there!"

Donald Anderson said that it did not take him long to figure out that Eric Pike was there to learn line work. "He wasn't there for the summer to make some extra money, he was there to learn what was going on," Anderson said. "Officially, I was classified as a groundman," said Eric Pike. "In reality, I was the grunt. And, after two weeks, I was the happiest grunt in the whole company. Dad had been right all along!"

"It seemed like to me, the harder you made it on him, the better he liked it," said Donald Anderson. "He didn't want any hand-outs; he wanted to be treated just like the rest of them. And that's the way he was treated by the other men on the crew: Roger Wright, the A/lineman, Chris Dockery, the C/lineman, and Everett "Ted" Ashburn, the equipment operator. Eric worked with our crew for two or three summers, and when he came in for Christmas and the other holidays, he always headed to our line crew and went to work with us wherever we were. Everybody enjoyed working with him, because he was easy to work with, and a lot of fun, really."

Anderson said that by the end of the last summer that Eric Pike worked for him, he had mastered his lineman's tasks. "All of us used to carry on with him that he was actually going to be far too good as 'lineman-material' to be wasted by working in the office, when we needed him in the field a lot worse than they did in there!"

When he finished college, Joe Pike wanted his son to come into the office. "This was something that I was just not ready for. I had pushed a pencil for the last four years, and I really wasn't interested in doing it again for a while." Pike said that he lasted in the office about two months. "Then, I made a deal with Dad. I told him that I had not finished what I had started. I was a low 'A' lineman when I came inside, but I wanted to become a full-fledged, confident 'A' lineman, and then I would come back."

Pike then joined Ronnie Newman's crew where he received a great deal of ribbing because he had turned in a company car, left an office and a computer, and had taken a pay cut to get there! "Dad had agreed to let me do it — reluctantly," Eric Pike explained, "but to this day, I appreciate that decision more than any other that he had to make about me while I was growing up." Joe Pike knew full well that in allowing Eric Pike to return to the field to learn more, he would be doing some of the most dangerous kinds of work that a lineman does. "By allowing me the opportunity, he really demonstrated his trust and belief in my abilities," said Eric Pike. "That's the way Dad is. He has so much confidence in all of us, that none of us ever want to disappoint him."

Pike added that it was during this extended time in the field that he learned many valuable lessons. "The last foreman whom I trained under was Gene Cook, one cantankerous old man, who was as rough and tough as they come, and my good friend." Pike and Cook first met when they were called in for storm trouble, and since the two of them were the first men to arrive at the show up, they were dispatched to work together on restoration that lasted most of the night.

"About 4:00 a.m., Gene was driving us down the interstate," Pike recalled. "The radio had been squawking for hours, and it was finally quiet, and the heater was on. My eyes began to get heavy, and I soon dozed off to sleep. Bang!!! I hit the windshield and the dashboard before I landed in the floor board of that truck." Pike said that his eyes must have been as big as saucers, when he looked up and saw Géne Cook still comfortably seated behind the steering wheel. "Gene, what happened? Did we have a wreck?" Pike asked. "Boy, if I am going to stay up all the damn night long," said Gene Cook, " then you are going to stay up all the damn night long." Eric Pike said that he was "boiling mad. But, I never fell asleep on storm trouble again!"

Pike clearly remembered that incident a few years later, when his general foreman assigned him to Gene Cook's crew as Cook's "A" lineman. "With my tools still in my hand, I looked at him and asked, "You aren't going to slam me into the windshield again, are you?"

"You ain't gonna fall asleep again, are you?" laughed Cook.

Pike's history demonstrates that almost all of Pike's management team and the company's line associates have progressed through experiences that were punctuated by tough, excellent guidance from men like Gene Cook. "This began one of the most special relationships I have ever had in my life," said Eric Pike. "Gene was the man who finally proved to me that I had attained the line-work knowledge I wanted." Pike said that Cook was a patient teacher who taught by giving him more and more responsibility. "He told me to plan and to do all the work that we had," said Pike, "and then he would 'grunt' for me. I learned a tremendous amount in a very short time from working this way. Gene taught me how to rig, and how to look for ways to let the equipment help me with the work. Mostly, by helping me plan the whole job, I learned more than just my part. But he told me that any time I was tired, or didn't know how to do something, to ask him and he would help."

Pike said that toward the end of the time that he and Gene Cook worked together, their crew was assigned a special weekend project. "When Cook told the general foreman that I should and could run the job," said Pike, "his confidence and referral finally assured me that, although I did not know it all — by any means — I had learned enough to be confident that I could plan and do most any kind of distribution work." Pike added that if Gene Cook didn't believe you could do the job, "he would quickly let you, and anyone else, know it!"

"Three years after I came out of the field, I had the great honor and sorrow to be a pall bearer at Gene's funeral," said

Eric Pike. "That was the first time I had lost anyone that had taught me so much, and I still miss him greatly today."

Working with Eric Pike in the Central Division is Ralph Anderson, who assumed additional duties as assistant vice-president in 1992. "Years ago, just before I finished my training at the lineman's school at High Rock Lake, Floyd gave me a ride home one weekend. On the way, he asked me to go to work for the company," Anderson said. "When I went to work as a groundman, September 19, 1953, I never dreamed that someday I'd have 700 people working under me." From groundman, Anderson progressed to a "C," "B" and "A" lineman, to a foreman, general foreman, supervisor, and area manager before becoming an officer. His experience covers the whole range of the company's work from building transmission lines, to high-line, hot-stick work and distribution jobs. He was even in charge of constructing a substation atop an R.J. Reynolds Tobacco Company plant in Winston-Salem. Anderson's area of responsibility is western North Carolina and all of South Carolina, but he has worked with the company from Maryland to Louisiana. "Without a doubt, of all the areas of the company that I've been involved with, the area that has seen the most growth has been the Carolinas — especially on the Duke Power Company system where, about fifteen years ago, we had fewer than 100 people, and now we have over 800. Also, in the Carolinas, we have hundreds more of our people who are working for co-ops and municipalities, and many others who are serving CP&L."

Also covering a part of the Carolinas and the Central Division is Audie Simmons, who became assistant-vice-president in 1992. "Although I officially came to work in the summer of '71," said Audie Simmons, I feel like I have been with the company all of my life, since my dad, Richard Simmons, has worked for the company forty-three years." Simmons remembered that as a child he looked forward to each Saturday morning when he came to the office and garage with his dad. "My job was to drop that envelope that held Dad's time sheets into the little slot in the door. I would then peek into that slot to see how many more envelopes were lying in the floor. Sometimes, Floyd would see me at the door and come out to pick at me and talk to Dad."

Audie Simmons said that he, like the children of so many others who came to the office and garage with their fathers, would play on all the equipment and pretend to be building power lines. "By the time I was twelve or thirteen, I was a pretty good climber, but I had nearly destroyed the pole in our back yard as I was learning," Simmons said. "When I put on Dad's 16-inch lineman-boots, climbing tools, and hard hat, I knew then that this was what I wanted to do when I got old enough — I just didn't know there were so many holes to dig by hand!"

Only forty-one years old, Audie Simmons already has twenty-two years of service with the Pike company, having started to work at the age of fifteen on a special worker's permit in Ashland. "I started as a groundman on Julian Green's crew — one hard working crew! My first day on the job, Julian took me to the top of a hill somewhere in Ashland and gave me a rock-bar and a set of snap-diggers. He said, "Start digging here, and don't stop digging until you are 7 feet deep!' " At the end of the day, Julian Green came up the hill and told Simmons that he had done a really good job, and that the crew could set *five* poles in that one hole. "I spent most of the summer with that rock bar and snap-digger in my hands, but after learning to shoot Kinepak, I could dig a pole hole in no time!"

Simmons said that after he became a full-time employee of Pike Electric on April 26, 1975, his training followed the conventional pattern. "At that time, I went to Coshocton, Ohio, and worked on Fred Anderson's crew. During the next twelve years, I spent most of my service on the AEP system. This was a great learning and training area for anyone involved in line work, because it entails nearly every aspect of the work — from doing it all by hand, to using the modern equipment, to making transformer connections, and all sorts of rigging. Most of all, the training teaches you to plan your work."

In commenting on his progression, Simmons added that on November 16, 1986, Claude Brinkley called him at home on a Sunday evening. "He asked me if I could meet him at Hardee's on Highway 601 in Mount Airy in 15 minutes. I said, ' Yes,' hung up, and wondered what I had done wrong." When he met Brinkley, he was told that he was to go to Raleigh the following morning and become the area supervisor of Pike's CP&L underground operations there. "This was another new opportunity, so I was ready to go," said Simmons. "It turned out to be more of a challenge than I had expected, but after a lot of hard work and long hours, we got it headed in the right direction, and I was soon promoted to area manager of the Central and Eastern Region of North Carolina for service to CP&L, Duke Power, REA and our municipal-systems customers.

*Richard Simmons and son Audie.*

Audie Simmons seemed to speak for everyone when he expressed his gratitude for the training that he has received from his predecessors. "Over the years, I have been very grateful for having the opportunity to work for, and with, some of the great planners and trainers in the company and to benefit from their years of experience. I feel that all of these fellows have helped to prepare me for the role that I play in the company today. Having started as a groundman, I progressed to lineman, foreman, general foreman, area supervisor, area manager, and my present position, assistant vice-president in our Central Division today. This has helped me to learn the business from the ground up. The learning process never stops, however, and today I am continuing to learn from our younger generation of management. Eric Pike has brought about some new ideas to our operation and has helped me to add some polish in my negotiating and dealing with our new generation of customers. I am thankful to all of those men who have been willing to share their knowledge and experience with me and others, but most of all, I am thankful to my dad. I feel that I have learned from the best."

Barny Ratliff said that he is frequently questioned about the many descendants of Pike employees that are now among the new generation of management as well as the hundreds of other employees whose parents are, or were, a part of Pike's personnel. "Several of our customers have adopted a nepotism rule that forbids the hiring of relatives, and I've had customers ask if Pike had a nepotism policy. I generally respond that we do not have a written policy, but our general rule is: 'We love it.'" He explained that this hiring practice has served the company well. "For the most part, these employees have a fierce loyalty to the company, with an attitude that approaches reverence. He said that the company has guarded against the potential downside to the practice by refusing to adopt the attitude of "owing" a job to any son or daughter. "If we hire this relative out of loyalty to the parent, at the expense of other more deserving, potential employees, we are not managing company for the future success that the company deserves."

Pike's history certainly demonstrates the positive aspect of the practice. "The Pike company has seen several sons and daughters follow in their father's footsteps over the years," said Audie Simmons. "The main reason that this has occurred is because of the strong dedication that the fathers have shown to the company and the opportunities that the company has provided. Personally, I am thankful for the knowledge and years of experience that Dad has passed on to me through his willingness to talk and build power lines at nights and on the weekends, as he answered all my many questions. He explained things based on experience in a way that I could understand, and he offered both positive and negative feedback to any situation at hand. This contributed to a great father-son relationship off the job, and a strong team effort while on the job."

In the Supply Department, Jim Draughn hired his brothers, Rudy and Gene Draughn, many years ago. "This is not the first time that we have worked together," Jim Draughn explained. "We were raised on a farm working together where we were taught, that if we respected one another, we would get along. Rudy worked for me on a transmission project in High Point before coming into the Supply Department." Draughn said that he and his brothers "play it strictly by the book. If something is wrong, it is wrong! No matter who it is. We have moral standards we live by. There is still a right and a wrong."

Jim Draughn said that when the positions came open in his

department, his brothers were available and ready for a job change. "When I told them about the openings, I told them that if they wanted a career at a job where they would be treated fairly, I didn't think they could find a better place. They've been here twelve or fourteen years, and there has been absolutely no friction from anyone." Draughn observed that everyone is treated the same. "Everyone is told when they are hired here, that we work as a family, a family of Pike employees."

The '90s ushered in changes in the Commercial Wiring Division also. Following Glen Pike's retirement after more than thirty years of service, his responsibilities were directed to Doug Porter who became assistant-vice-president in 1994. With a background in both management and engineering, Porter had been general manager of the L.S. Starrett Company in Mount Airy for nearly thirty years before coming to Pike. He followed Glen Pike's path and quickly studied for and obtained his unlimited electrical license. Ever aware of the legal aspects that are tied to every job, Porter also went to Construction Law classes as well. Meanwhile, he worked to refit the HVAC Shop early on and to streamline its production, with positive results. At the same time, Porter began to move the division into the "computer age." The Commercial Wiring personnel quickly began to employ "PC's", expediting many of the department's tasks. During the early days on the job, Porter worked with Joe Reid to integrate the Pike North facility into the company's main system.

Assistant-vice-president, Jim Benfield, an employee with Pike, since his graduation from North Carolina state in 1985, also has a background in engineering and management and is the general manager of the Commercial Wiring Division. Benfield's first assignment with Pike was in the Commercial Wiring Division, but through the years, he has gained experience in other aspects of the company from line work to project estimator. Before assuming his new duties, Benfield worked was manager of operations for Pike's Central Division.

"The nucleus of the management team has experienced a major change in the last few years," said Joe Pike. "Yet during this time, we have remained profitable, more than doubled the size of the company, and reacted to the many major changes that our customers required as a result of their own reorganizations. We were able to make these transitions successfully because our retiring officers left their divisions in excellent running order, and they prepared an excellent group of new managers to carry on."

These new managers have not only had field experience, but they are very comfortable with a high level of technological adaptations. They are eager to get and use the latest technology. "Even more important," Joe Pike continued, "these new managers bring an appreciation of the skills and the values of those who have come before them, along with an awareness of the importance for continuing the traditions of hard work, safety, and integrity that is their heritage here."

Pike's new managers, like their predecessors, have tremendous talent and capability, and they share with their predecessors a willingness to take tremendous risks. It is interesting to note, that as Pike begins a new era, the management team is approximately the same average age as the original team, when it began the company's first fifty years. Pike's future is in good hands.

*Pike's Management Team:*
*L-R – Kelly Surrott, Audie Simmons, John Brewington, Reg Banner, Eric Pike, JoePike, Barny Ratliff, Jeff Collins, Ralph Anderson, John Merritt, Doug Porter, Zack Blackmon, Jr., Jerry Boatright, Johnny Mullins. Officers – Not Pictured: Jim Benfield*

*L-R – Reg Banner, Opal Pike, Glen Pike, Emmett Forrest, Claude Brinkley, Joe B. Pike, Haywood Merritt, Zack Blackmon, Roy Thomas, Barny Ratliff, Eric Pike*
*The Board of Directors of Floyd S. Pike Electrical Contractor, Inc.*

# 38

# Storms of the '90s: From Andrew to Opal, and Ice in Between

Hurricane Hugo still ranks as the storm of all storms for the Pike company. Although there were no natural disasters in the 1990s that required a company-wide effort, hurricanes Andrew and Opal, and the ice storms that occurred in between, created their own challenges.

Hurricane Andrew ravaged south Florida in 1992. Although Andrew's rage left total devastation in its wake, the storm's path touched only a narrow area. Over 300 of Pike's personnel were directly involved in the restoration effort as Pike's crews were mobilized, and caravans of yellow trucks headed south where they worked under armed guards to protect them from snipers and their equipment from theft.

Following the immediate repairs, Florida Power and Light requested that a number of the crews remain for clean-up work. By that time, many of Pike's employees had already been away from their homes and families for over three weeks. However, other members of Pike's personnel became available for rotation, and, moreover they were willing to work through the Christmas holiday. To rotate people was one thing, to rotate a convoy of over fifty trucks — already there and working — was another matter. Equipment transferal to the many points from which it had been removed and acquiring flight schedules for hundreds of people created logistical problems that required innovation.

Rather than return the trucks, the decision was made to remove and return only the customized equipment that was installed on the vehicles. For example, the trucks that had been driven by crews who had been released from Duke Power, were equipped with the power company's specialized radios. Loads of equipment were hauled back to normal work sites in the Carolinas and Georgia aboard the company plane.

Some crews remained, and others returned home on commercial flights. Pike had personnel waiting at the airport to meet the returning crewmen and to provide ground transportation. The excitement at the airport resembled that of a military squadron returning from deployment — with wives, children, girlfriends, and other family members eagerly greeting the returning Pike employees.

The winters of 1993 and '94 caused major restoration efforts for Pike. The infamous Blizzard of '93 struck much of Pike's service area. The storm began on March 20 and caused heavy damage to the systems of the north Georgia co-ops. Pike's customers in the central and western areas of the Carolinas were also hard hit, and customers in eastern Tennessee and western Virginia were without power. At the height of the storm effort, the Pike company had over 1,000 linemen working to restore power over a six-state area.

Pike's restoration effort suffered a unique twist in this blizzard. The company's usual rapid response was impaired, due to the huge accumulation of snow which closed all of the major highways in the region for hours and delayed restoration.

When Interstates 77 and 85 in North Carolina, I75 and I85 north of Atlanta, and I-40 on both sides of Asheville were reopened, the yellow trucks were, literally, leading the way. And with Pike crews from every region involved in the effort, power was restored in the affected areas in only seven days.

The following year, the same areas were affected by two back-to-back snow and ice storms. The first storm started on December 21, 1993 and ended late on Christmas Day. Pike crews completed most of the restoration within forty-eight hours. However, in Pike's Central Division, over twenty-five crews, from as far away as Greenville, South Carolina, worked on through Christmas Day to help restore power for Surry - Yadkin EMC customer at Dobson. "This type of dedication is why Pike Electric and its employees have the reputation that they do," said Eric Pike. The second storm affected the same areas but was handled by local crews working only a few days.

Usually ice occurs only in relatively small geographical areas, but this was not the case in February 1994 when moist gulf-coast rain mixed with a dip in the jet stream and created an extremely long band of ice storms that began in Texas and ended in Maryland. Pike crews in Alabama and Georgia responded immediately as the storm ravaged areas in both of those states early on. By the second day of this brutal weather, portions of the Carolinas and Virginia, as well as southern Maryland — directly in Pike's service areas — were hard hit. Other areas in southwestern Virginia, central Tennessee, western Mississippi and Arkansas were also devastated by ice.

Philip Combs, area supervisor, Southern Division, lead the first group of Pike crews from Georgia all the way to Arkansas. When he arrived, he called in to report that even old timers there could not believe the destruction. One power company official said the area looked like it had been hit by a neutron bomb.

"This series of storms involved the best coordination and movement of crews and equipment Pike Electric has ever accomplished," said Eric Pike. Crews that had left home and headed to Virginia, found themselves continuing on to Mississippi or Tennessee where the work was extremely difficult. "Many crews had just completed work in below-freezing temperatures in Virginia," explained Eric Pike, "and then found themselves in waist-deep mud, just days later, in Mississippi."

And the damage in Mississippi was as severe as any that Pike's long-time employees had ever seen. The Mississippi Delta is primarily devoid of trees — and falling trees usually cause most of the damage in normal winter storms. However, it was the sheer weight of the tremendous amount of ice that the area received which left transmission and distribution lines twisted and tangled on the ground. Substation structures even collapsed under the frozen load.

Bo Lankford, area supervisor from Pike's Central Division, lead a second convoy of Pike employees from North Carolina into the Mississippi area. Nearly 100 men strong, this group had just come from restoring power on the Nantahala Power & Light system in the North Carolina mountains, where they had encountered snow drifts several feet high. These men didn't think Mississippi would be too bad.

When Pike's crews neared the storm-damaged area and spotted the first broken pole, one lineman broke the silence on the company radio and announced to the listeners following in the convoy of yellow trucks, "Boys, we won't be here too long." A few more miles down the road, the damage got a little worse, and a different voice came across the radio. "I'll say it will take about a week." As the group pulled into Greenwood, Mississippi, a lone voice came across the radio. "Boys, we're gonna be here a while."

Pike employees were in Mississippi and Tennessee for three weeks on storm work. Later another group went back to Mississippi to help with clean up. All told, Pike Electric had over 500 people working storm restoration in Mississippi, and, at the same time, there were still over 500 working storm work in Virginia. When all of the crews finally returned to their homes by mid March, some employees had been away for nearly a month.

"Pike's assistance to the Nashville Electric Service during this storm was the company's first opportunity to show this customer our capabilities," said Barny Ratliff, "and we made a very good impression under adverse conditions that has paid dividends, as NES remains a regular customer today."

Although, the restoration required driving long hours to work in very rough conditions, Pike employees had no major personal injury accidents and only two vehicle accidents," said Eric Pike. "This really proves that Pike's safety and training pays off."

Hurricane Opal hit Pike's customers during the first week of September, 1995. The storm had crossed Florida causing little damage. However, when "Opal" reached the Gulf of Mexico, the hurricane regained strength, and then came through the South's back door, charged into southern Alabama, northern Georgia, western North Carolina, and on into Virginia.

Hurricane Opal brought tremendously high winds and a good deal of rain much further north than anyone expected, and Pike immediately began mobilizing its forces. "Almost all of our crews in the Southern Region headed for Alabama and north Georgia," said Eric Pike. "Although these were the two areas that were the hardest hit, calls quickly came for many additional crews in western North Carolina." The company had over 1,200 linemen working at the height of the restoration effort, and it was almost ten days before power was fully restored in the affected areas and Pike's personnel could return home.

On a personal note, that is in no way intended to lessen the seriousness of the storm; around the home office in Mount Airy, almost everyone — especially Pike's own "Opal" — considered the National Weather Service's choice of names for this storm to be highly appropriate and timely. Hurricane Opal — "the storm" of our 50th anniversary year.

*Pike's own "Opal".*

Page 4

*A Letter From . . .*

Landwood Acres Bed & Breakfast

Rt. 1 Box 190, Arlington, VA 22922

February 21, 1994

To the Editor:

In the year I have operated a bed and breakfast, I have had a diversity of guests, but nothing to equal the 8-member crew of linemen brought in by Appalachian Power from North Carolina. They were billeted in the bed and breakfast for two nights following the ice storm. They arrived at 11:45 p.m. on Thursday night, one week after the storm struck, exhausted, aching and very, very homesick. The night before, they had not found accommodations until 3:00 a.m. in Lynchburg, and were back on the lines at 9:00 a.m. with a 4:00 p.m. break for dinner. Before arriving in Nelson County on Monday or Tuesday (they really could not remember which day), they had worked a line in South Carolina. I never realized to what extent a hot bowl of homemade soup, a pot of stew with dumplings, salad, and chocolate cake could be appreciated. I only have four rooms, but they did not complain about the inconvenience of two to a bed — they were just happy to be "home"! Morning call was for six with breakfast at seven, and they started arriving in the kitchen about 6:30 for a first cup of coffee.

There were no laggards. They were gone by 7:30. The second night, they came in a little earlier — at 9:45 — and opened the kitchen door with calls of "Mom, we're home!"

These truly "unsung heroes" bring back our electrical power with very little appreciation and understanding of the dangers they face and sacrifices they make traveling in their equipment from one line to another in obscure areas with few conveniences, or the number of hours they work without rest. The most pain I saw was during the telling of the total lack of understanding of one woman in a restaurant who thought it was a disgrace they were eating in the middle of the afternoon and who was disgusted with seeing their trucks in restaurant areas at all hours of the day while people were without the convenience of power. These are not nine-to-five men with lunch from twelve-to-one. They work in remote areas and eat when they can, not when they would like, and frequently work for stretches of eighteen or more hours in conditions the average person could not conceive of. The second morning when they left, they had to return to an area which their trucks could not reach (the reason for stopping so early the night before!) and equipment had to be hauled in manually.

I only wish I could relay some of the incredible stories they told. They are very dedicated family men who spoke with love and tenderness for their wives and children — even showed pictures. Under all the gruffness, I found warmth and politeness that exceeded any that I have ever encountered. When they left here Saturday morning, they had only one hope — to finish and be home that night even though it was not going to be for more than a night or two. Kentucky and Tennessee still had remote areas where it was anticipated power would be out for three weeks, and that was their next assignment.

We all owe so many, many of these linemen our "thanks." And I know the next time we have a "natural disaster," thanks would be a welcome change from the usual complaints. When they were leaving, one of the men said that staying here "changed their attitude — for it was a touch of home so far away." How very kind of him. Thanks, Steve, Chris, Jerry, Coleman, Reid, John H., John S., and Roger. It was my honor to have you here!

Lynn Holmes
Landwood Acres B&B
Colleen

*Our crews in Nashville worked or checked 6,189 call-outs in 7 days, or an average of 25 calls per truck per day.*

Eleanor M. Atkinson
205 South Bolivar
Cleveland, MS 38732

February 23, 1994

Floyd S. Pike Electrical Contractor, Inc.
P.O. Box 868
Mt. Airy, NC 27030

Dear Sir:

The words "thank you" seem so little to say to your employees who have come to the Mississippi Delta to help following the ice storm of February 10th. I wish there were some way we could personally thank each of them and let them know how much their being here for us means to the residents of Bolivar and Washington Counties.

Shortly after the ice storm I saw convoys of trucks coming to our aid. It may sound ridiculous, but my heart was filled with gratitude to know so many would travel so far to help in our time of need. I work in Greenville and live in Cleveland; every day when I commute I see your dedicated workers tirelessly putting our utilities back in working order. Thank you!

Many citizens of the Delta are displaying white ribbons to show our gratitude, and we hope the workers notice and understand what we are trying to say. In case your crews do not know how thankful we are for their help, I ask that you pass our thanks on to them and to their families who have been separated from their loved ones during this difficult time.

Sincerely,

Eleanor M. Atkinson

Eleanor M. Atkinson

Mr. Roy Thomas, Vice President
Pike Electrical Contractor
P.O. Box 868
Mount Airy, N.C. 27030

Dear Mr. Thomas:

I want to express my gratitude for the wonderful assistance that your crews gave to the Oneonta District of Alabama Power Company during the Blizzard of '93.

Your personnel worked under some very difficult circumstances and they also worked some very exhausting hours and did so with dedication, cooperation and professionalism. They always arrived ready to give their best and it was truly a pleasure to work with them. I have received many accolades about the Pike crews from not only our customers but our own personnel as well. They helped us to accomplish a very demanding, and at times seemingly impossible task in a minimum of time.

Please convey my thanks, along with the thanks of all the employees of the Oneonta District, to all the members of the Don Combs' crew for the long hours and hard work put forth for our customers. They have made some new friends and are indeed a credit to your company.

With much thankfulness,

James P. Henderson
Manager-Operations

7185 Southaven Circle
Southaven, MS. 3867
MARCH 7, 1994

Dear Floyd S. Pike Electrical,

My name is Chris Hayden. I Go To Hope P. Sullivan School. I'm in Mrs. Hollis's HomeRoom. Thank you for Helping us out Down There. My Family and I appreciates all of yall for Helping us From The "Ice Storm of 94". I Couldn't of gone an other Day without heat, lights, Bathroom, or my Super-nintendo. Thank you a lot!

Sincerly,

Chris Hayden

*Duke Power was Pike Electric's first customer.*
*We are very proud to have worked continuously*
*with Pike over the last fifty years.*
*Duke takes pride in its reputation*
*for providing reliable service and*
*customer satisfaction at competitive costs.*
*Pike Electric has played a critical part*
*in our ability to achieve these measures of success.*

*—Jim Hicks*
*Senior vice-president, customer operations*
*Duke Power Company*

39

# Single Supplier Award: Duke Power Contract

The first four months of 1992 were anxious and potentially troubling times for the company. The normal seasonal reduction in the company's workload during these months was compounded by a recession. "Our employment levels were around 2,000 people at the end of 1991," reported Barny Ratliff. "By the end of April, our employment numbers had dropped to 1,637." Crews were being released by customers with ongoing contracts, and work that was bid during this period was being awarded to many of Pike's competitors at or below the company's cost. The market was extremely tight.

Then the company received the news that Duke Power intended to competitively bid its distribution contract. At the time, approximately one-fourth of Pike's work force was at work on the Duke Power System. "The loss of these crews would have caused us to make drastic changes in the company," said Ratliff. "And with economic conditions as they were in our region, we would not have been able to obtain work in other locations to place even a part of these people, if we were unsuccessful in getting the bid." The entire company would be adversely impacted, if it could not at least maintain the numbers of crews that were then serving the Duke account.

In late February '92, Joe Pike, Barny Ratliff and Zack Blackmon, Sr., attended the pre-bid meeting in Charlotte. "We had heard that over forty contractors had completed a bidder-qualification form in order to be allowed to bid Duke's contract," said Ratliff, "but only fourteen contractors had survived the qualification process, and they were all present at the meeting — seven were already working on the Duke system — and the other seven were potentially new contractors." The bid package implied that Duke wanted to award contracts on a regional basis, although the possibility of a single contract award was mentioned. The regional aspect of the contract was important to Pike, because the company already had varying amounts of work in each of Duke Power's three regions.

"We had less than three weeks to prepare a proposal according Duke Power's format and their invitation to be creative," Ratliff remembered. Pike went to work. For several days and nights, Joe Pike hardly left his computer. A realistic decision had to be made concerning a feasible volume of work. "The prospect of having no work on the Duke system was always a possibility that influenced our thinking as well," said Barny Ratliff.

Considering the tight market for contractor services, Pike knew that the successful bidder would submit prices well below the current rate for work on the Duke system. "We could either reduce prices and rely on our reputation for quality service and hope to retain one to two regions," explained Ratliff, "or we could reduce prices even more dramatically and give Duke the opportunity of a single award for their entire system. The decision was made to format our proposal to cover both options."

The bid was submitted on March 12 to the Duke Power Commodity Team. The team examined the bids and graded contractor evaluations, ranking them in various categories of service and capabilities. Pike was at the top of each category except one. "We never did find out in which category we were ranked second," said Ratliff. Only after the evaluations were completed, were the bids opened. "The group was shocked to discover that not only was Pike the top-rated contractor, but the Pike company had also submitted the best price."

The evaluation process took about one week, and unbeknownst to Pike at the time, the team then spent the following two weeks convincing Duke's management to make a single award. "We had no idea what was going on during the evaluations in this three-week period," Ratliff remembered. "It was an anxious period of time for everyone in the Pike organization."

Finally, on Monday, March 30, Jim Hicks, senior vice-president, customer operations, called at 8:00 a.m. He and other Duke personnel wanted to meet with Pike's management to discuss the contract at 11:00 that morning. Jim Hicks, Donna Clutz, and Jerry Chapman came to Mount Airy and met with Joe Pike, Barny Ratliff and Zack Blackmon, Sr. Hicks briefly outlined the process that Duke had employed in making its decision, and then he presented the single supplier offer. Joe Pike readily accepted. "We were told that of all the contractors considered, Pike was the only company that Duke would consider for a single-source supplier," said Ratliff. This award has remained a source of pride for everyone in the Pike organization. Shortly after the contract was awarded, Bill Coley, senior executive with Duke Power commented:

> The Pike organization has established itself on the basis of delivering quality service, being responsive to its customers, and having the highest ethical standards in their business practices. That's the kind of thing that those of us

> in this business look for, when we look for partners. We also look for people who understand that their key to success in is helping us, as their partner and their customer, be successful.

In its proposal, Pike had indicated the company's ability to obtain the equipment necessary to move from a level of 185 crews to the total of 431 crews that were required. Equipment purchase began immediately. Pike then began displacing its competitors in May '92, and by mid-June, the company was the major power line contractor on the job, and crew levels had stabilized at 374. Pike had successfully started over 190 crews in only six weeks, more than doubling its Duke operation.

Duke Power enjoyed a significant price reduction in costs per crew and a 13 percent reduction in crew numbers. Only 374 Pike crews were required to replace the 431 crews that were in place prior to the bid.

At the end of that year, Zack Blackmon retired, and when Eric Pike took charge of the Central Division, he inherited an exciting situation that was really in flux. The bold move made by both Duke and Pike to agree to a single source contract in '92 came at a time of upheaval at Duke Power Company. Duke reorganized its Customer Services group shortly after the contract award, and almost an entirely new slate of players appeared on the scene — all with different ideas on exactly how this contract was to proceed.

Problems began to occur later in '92, when crew levels dropped to below 320, more than 110 crews below the anticipated initial level. However, the major problems occurred late in '93 and early '94, when Duke made changes in its approach to overhead distribution construction. This change reduced Pike's overhead forces by over 100 crews. The opportunity for the upward movement of the company's work force disappeared, and the wage-rate adjustment shifted from overall positive to predominately negative.

By the end of the contract's second year, Duke had reorganized and downsized, and the power company was entering into another major reorganization. Pike personnel were then working with thirty-two different key contacts there, ranging from executive vice-presidents to construction and operation supervisors. And following the second reorganization, none of the original thirty-two people were in the same positions: nineteen were gone, and of the remaining thirteen who had been involved with Pike's personnel and its contract in the beginning, many were assigned to different positions and no longer had direct contact with Pike.

Toward the end of Pike's fiftieth year, it was obvious that the structure and organization of Duke Power was totally different from what it was at the time of the original bid. Nevertheless, Eric Pike worked tirelessly and tenaciously, and made numerous trips to Charlotte to work with Duke's personnel to address the necessary contract amendments in the wage-rate inequities and work procedures that had occurred as a result of the changes in the utility's approach to line construction.

With integrity, commitment, and pride, Floyd S. Pike Electrical Contractor, Inc., continues to work for its very first customer — a relationship that has lasted for over fifty years now.

*The great thing about it is that it is not over.*
*With the imagination and sound business judgment*
*of the new generation of leadership that we have,*
*what develops in the next fifty years*
*may make the accomplishments of the last fifty look small.*
*It will really be something to see.*
*Too bad we old fellows won't be around*
*to participate in the show.*
*Don't get me wrong. I'm sure not ready to hang it up yet.*
*I consider myself to be one of the luckiest people around.*
*Very few people are allowed to do what they enjoy doing*
*for what amounts to their entire working career.*
*In closing, let me say, that on a clear day the view is*
*breathtaking from the top of a 350-foot steel pole in the middle*
*of the Potomac River.*
*Everyone should see it. I feel like I stood there*
*and watched a two-lane highway dream*
*develop into an eight-lane expressway.*
*Reality.*

*—Charles M. "Droopy" King*

40

# Keep On Climbing

"There are many images that come to mind," said Robert Chappell, "when I reflect on a twenty-five year relationship with Floyd S. Pike Electrical Contractor, Inc., — Floyd Pike and his enthusiasm for the business. Old No.1. The stockholder meetings. And the yellow trucks spotted in so many unexpected places; and today's Pike company — the result of many events and the efforts of countless people who have contributed over the years."

The next fifty years will be years of fast, exciting change that will present tremendous challenge, a word, you may remember, that Floyd Pike defined as *opportunity*.

As rapidly as technology is developing, it is not inconceivable, for instance, that Pike may soon see on-line computers in line trucks with integrated Global Position Mapping. New technology will require smart people. Pike is ready. "Anyone who can confidently change out a junction pole or go into a *live*-front underground cabinet and switch feeds, should have no trouble with any new technology," said Joe Pike. "And this company has a work force filled with people who are capable, ready, and eager to meet any change that comes, and any challenge that a change may bring. *Challenge* still means *opportunity* around here."

The management team is poised for change. "In the past," said Claude Brinkley, "management was made easier by the fact that we had a more direct route to our customers' decision-makers. We had an outstanding working relationship with the people that we worked with, and they were able to understand the problems that we had. In many cases, we worked with the same decision-maker for decades. With downsizing and reorganization, and the increased size of the companies that we deal with, often there are *many* people involved in making decisions that concern us, and sometimes they really do not understand the work that we do."

Pike's recent history has shown that no matter what trends come and go in today's business environment, Pike just keeps on doing whatever is required to get the job done.

"The days of 'peach brandy and pistachios,' when many problems in the company's business were solved almost before they developed, and a supplier was chosen according to a customer's own criteria, rather than by the cheapest service offered, seemed to disappear for a time," said Joe Pike. "However, as more and more of our company's customers become financially independent entities, we are seeing the trend once again returning, when a company will chose its supplier based on quality service and matching values."

"This company has gone through some tough times in the past," said Claude Brinkley, "and it would be unrealistic not to think that somewhere down the road there may be other tough times. However, we've always had and will always have a management team that can cope with any kind of problems."

"No doubt our industry as a whole is on the threshold of great change," said Joe Pike. "Nevertheless, electrical service will still need to be delivered to all of today's users, as well as to new users as they develop. The source of the structure of the delivery, and the choices for the user, will, in all likelihood, change; but for a considerable time, the need for a delivery system that conveys energy from its source to the final user will remain. This a market for our company to supply, for many years to come."

Considering the international market of the utility industry today, Claude Brinkley's hopes for the future are not unfounded. "I would really like for the management team that we have now and the management team that we will have in twenty-five, thirty years or more, to look far ahead into the future. I would like for them to come to work every morning visualizing *World Headquarters of Pike Electric* on the big granite sign out front."

The Pike company's immediate future includes new facilities. Plans are underway to consolidate all of the separate operations in Mount Airy at one location. Initially, a new headquarters building and a hangar will be built on a tract of more than one hundred acres, adjoining the Mount Airy-Surry County Airport. Additional facilities for garages, commercial wiring and supply and testing will follow. After the decade of struggle that Floyd Pike led to build and then maintain the airport, the company's new location seems, if not destined, at least appropriate.

"The reason Pike is celebrating its fiftieth Anniversary is due to the commitment that Floyd Pike made in the very beginning to quality work, customer service, and fair dealing," said Douglas W. Booth, former president of Duke Power Company. "As a result, he built a successful, profitable business which provided jobs to employees, profit to owners, and satisfaction to clients."

In compiling its history, the Pike company has looked back, only briefly, to see where it has been and to measure its progress. However, Floyd S. Pike Electrical Contractor, Inc., is not, in the words of its founder, a "look-back" organization. "We, at Pike, are rightfully proud of our past," said Joe Pike, "and we are grateful to all of the thousands of people who are, and have been, a part of our history. We begin our next fifty years dedicated to the legacy that we have inherited. With our integrity and determination, we are committed to meet the future with success."

Climbing to the top has brought Pike Electric to a "mountaintop moment" — a time to lookout and see anew our opportunities and to rededicate ourselves to the tasks before us," echo the words of Floyd Pike. "There never has been a need for the word, contentment — in fact, we are just beginning to grow!"

ELECTRICAL CONTRACTOR

# EPILOGUE

As we begin another fifty years, I wish continued success to each person who has been a part of the accomplishments made by Floyd S. Pike Electrical Contractor, Inc. It is gratifying to me to see that our company continues with the same values and goals that have been in place since the beginning, and that the company continues to provide opportunities for our employees.

When Floyd and I began, our goals were the same as yours today — to buy a home, to educated our children, to share our success, to prepare for our retirement. I am happy that, with God's help, and yours, the climb continues.

— Opal F. Pike

*Opal Pike – 1995*

*Opal Pike with old Number One on Pike's 50th Birthday, Dec. 3, 1995. They accompanied the prize winning Pike Electric float in the Mt. Airy Christmas Parade.*

*Opal Pike cuts Pike's 50th anniversary cake.*

# Endnotes

[1]Joe Maynor, *Duke Power: The First 75 Years* (Charlotte: Duke Power, 1979) 92
[2]Maynor, 92
[3]Maynor, 94
[4]Jimmy Webb, "Wichita Lineman," Los Angeles, Canapy Music, 1968
[5]Maynor, 92
[6]Maynor, 129, 127, 132
[7]Maynor, 112
[8]Letter and memorandum received 26 March 1993 from. B.B. Parker, former president, Duke Power Company.
[9]W.B. McGuire, Letter to Opal Pike, 19 April, 1993
[10]W.B. McGuire, Telegram, 19 April 1963
[11]Photograph and dollar courtesy. B.B. Parker.
[12]Letter, B.B. Parker
[13]Journal and Sentinel, 4 April, 1965. "Banker Ed Duncan: Symbol of Northwest N.C. Growth", Sec. D, 1
[14]Eric Hoffer, *Reflections on the Human Condition,* (New York: Harper & Row, Publishers, 1973) aph. 174
[15]John Banner, Jr., Interview, February 17, 1996.
[16]Maynor, p 157
[17]Sam Walter Foss, "The House by the side of the Road" Based on lines from Homer: "He was a friend to man, and he lived in a house by the side of the road."
[18]Richard E. Maxey, chairman Federated Bancorp, Inc., Loda, Illinois, Letter 20 March, 1996.
[19]John W. Vaughan, former president, Appalachian Power Company, Roanoke, Virginia, Letter 25 March, 1996.
[20]Douglas W. Booth, former president, Duke Power Company, Charlotte, North Carolina, Letter, 8 March, 1993.

PIKE

Since 1945

50 YEARS